Ground Cover Fire Fighting for Structural Firefighters

First Edition

By Thomas E. Richter

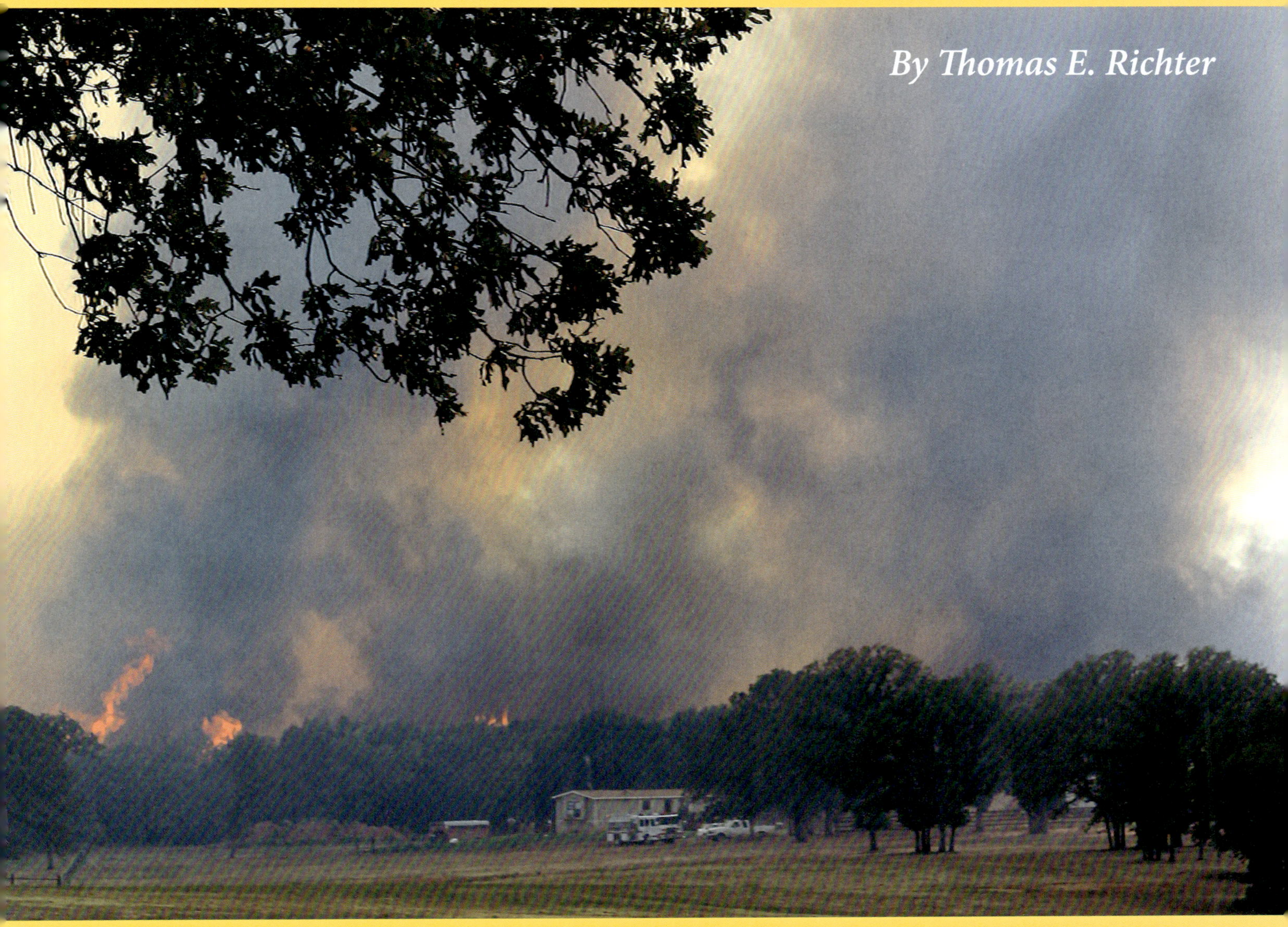

RECYCLABLE

Published by
Fire Protection Publications
Oklahoma State University
Stillwater, Oklahoma

The following staff members at OSU Fire Protection Publications contributed to the final publishing of the manual:

Senior Editor/Project Manager Cindy Brakhage
Senior Editor Jeff Fortney
Graphic Designer Errick Braggs
Associate Director Mike Wieder

Copyright © 2018 by the Board of Regents, Oklahoma State University.

All Rights reserved. No part of this publication may be reproduced without prior written permission from the publisher.

ISBN 978-0-87939-671-8
Library of Congress Control Number: 2018935433
First Edition
First Printing, April, 2018
Printed in the United States of America

Oklahoma State University in compliance with Title VI of the Civil Rights Act of 1964 and Title IX of the Educational Amendments of 1972 (Higher Education Act) does not discriminate on the basis of race, color, national origin or sex in any of its policies, practices or procedures. This provision includes but is not limited to admissions, employment, financial aid and educational services.

Contents

About this Manual x

Introduction xi

Chapter 1 Ground Cover Fire Behavior 3
Types of Ground Cover Fires 4
    Ground Fires 5
    Surface Fires 5
    Crown Fires 5
Ground Cover Fire Behavior 5
    Ground Cover Fuels 5
    Weather 6
    Wind 6
        General Winds 7
        Local Winds 7
    Air Temperature 8
    Relative Humidity 8
    Precipitation 9
    Atmospheric Stability 9
    Topography 11
        Aspect 12
        Slope 13
        Terrain 13
Firelines 14
Extreme Fire Behavior Characteristics 15
Key Terms 16
References 17

Chapter 2 Firefighter Safety and Survival 21
Operational Leadership 21
Risk Management 22
Standard Firefighting Orders 23
Turn Down (Refusing Risk) 26
Eighteen Watchout Situations 27
    Lookouts, Communications, Escape Routes, and Safety Zones
    (LCES) 31
        Lookouts 31
        Communications 32
        Escape Routes 32
        Safety Zones 32
Personal Protective Equipment 35
    Structural Gear 35
    Wildland Gear 36
        Wearing PPE on the Fireline 36
        Fire Shelter Basics 37

How the Fire Shelter Works	*38*
Determining Fire Shelter Location	*39*
Agency Policy	40

Rules of Engagement **40**
   Situational Awareness 40
   Look Up, Look Down, Look Around..41
   Look Up 41
   Look Down 42
   Look Around 42
      Avoiding Lightning *42*
      30/30 Rule *43*

Common Denominators of Fire Behavior on Tragedy Fires **43**

Wildland Fire Apparatus Safety **44**
   General Guidelines 44
   Off-Road Guidelines 44
      Capabilities and Limitations *45*
      Hoseline Safety Guidelines *45*
      Cautions in Off-Road Apparatus Operation *45*
   Personnel Transport 46

Heavy Equipment Operations **47**

Heavy Equipment Hazards **48**

Seeing and Being Seen **48**

Heavy Equipment Operations Safety **48**
      Hand Tool Safety *49*
      Radio Communication *49*
      Hand Signals *50*
      Safety Tips for Firefighters...*50*
      Heavy Equipment and Underground Utilities *50*

Hazardous Materials Situations **51**

Operations Near Electrical Power Lines **53**

Snag Safety **54**

Key Terms **55**

References **55**

Chapter 3 National Incident Management System —The Incident Command System (NIMS-ICS) 59

NIMS-ICS Organizational Functions **60**
   Command Section 61
   Incident Commander (IC) 62
      Deputy Incident Commander *63*
      Incident Safety Officer (ISO) *63*
      Public Information Officer (PIO) *64*
      Liaison Officer *64*
      Incident Command Post (ICP) *64*
   Operations Section — Operations Section Chief 64
   Planning Section — Planning Section Chief 65
   Logistics Section — Logistics Section Chief 66

Finance/Administration Section — Finance/Administration Section Chief 67
Unified Command 68
Incident Management 68
    Delegation of Authority 68
    Organizational Principles 69
        Scalar Structure 69
        Unity of Command 70
        Span of Control 70
        Division of Labor 70
    Management by Objectives 71
        Size-Up 71
        Incident Strategic Goals and Tactical Objectives 72
    Resource Management 72
        Resource Terminology 72
        Aid Agreements 73
        Anticipating Resource Needs 74
    Incident Communications 74
        Formal Communications 75
        Informal Communications 75
        Briefings 75
    Incident Action Plan (IAP) 76
        Verbal IAP 77
        Assigned Incident Tasks 77
        NIMS IAP Planning Process 77
Key Terms 78
References 78

Chapter 4 Strategy and Tactics 81
Parts of a Ground Cover Fire 81
    Origin 82
    Head 82
    Fingers 82
    Perimeter 82
    Heel 82
    Flanks 82
    Spot Fires 82
    Islands 82
    Slopover 83
    Green 83
    Black 83
Size-Up 83
Forming an Incident Action Plan (IAP) 85
Fire Control Strategies 86
    Direct Attack 88
    Indirect Attack 89
    Locating and Developing the Fireline 90
    Anchor Points 91

   Fireline Width 91
   Fireline Construction 92
   Line Construction with Mechanized Equipment 96
Key Terms **97**
References **98**

Chapter 5 Ground Cover Engine Operations 103
Structural Apparatus Used on Ground Cover Fires **103**
Resource Typing-Incident Command System (NIMS-ICS) ... **105**
 Resource Typing 106
 Wildland Engine Types 107
  Type 3 107
  Type 4 107
  Type 5 108
  Type 6 108
  Type 7 108
Fire Control Tactics **108**
 Flank Attack 109
 Pincer Attack 109
 Frontal Attack 110
 Mobile Attack 110
 Tandem Attack 112
  Two Engines 112
  Hand Crews 113
 Hotspotting 113
 Indirect Attack 114
  Advantages of an Indirect Attack 115
  Disadvantages of an Indirect Attack 115
Hose Lays **115**
 Booster Line 116
 Progressive Hose Lay 117
Ground Cover/Urban Interface Operations **118**
 Command, Control, and Accountability 119
  Strike Team 120
  Task Force 120
 Ingress and Egress 120
  Residents and the Public 121
  Evacuation 121
  Routing Traffic and Establishing Access 121
 Structure Triage 122
  Greatest Potential Threat 122
  Probable Threat 122
  Factors Affecting Triage 123
  Consider All Factors 127
  When Structures Cannot Be Saved 127
Structure and Site Preparation **128**
 Structure Protection: Lessons Learned 128
 The Structure 129

  On-Site Resources 129
   Locate Water Sources *129*
   Adjacent Resources *129*
 Clearance Around Structures 130
 Removing and Trimming Fuels 130
Fireline Construction **130**
 Intermediate Fuels 131
 Yard Accumulation 131
 Flammable and Explosive Hazards 132
Structural Exterior and Interior Preparations **132**
Private Vehicles **133**
Pets and Livestock **134**
Pretreatment of Structures **134**
 Sprinkler Systems 134
 Class A Foam 134
 Fire Gel 134
 Structure Wrap 134
Structure Protection Tactics **134**
Working Hoselines **135**
Nozzles **137**
Confronting the Fire at the Structure **137**
 Spotting Zone 137
 Full Containment 138
 Partial Containment 138
 No Containment Possible 138
 Fighting Roof Fires 138
Water and Foam Use **139**
 Water Supply 139
  Water Application *139*
  Wetting Down with Water *139*
 Reducing the Heat Buildup 140
  Duration of the Heat Wave *140*
  Peak Heat Wave Tactics *140*
 Class A Foam 140
  Properties of Foam *141*
  Types of Foam *141*
 Structure Treatment 141
Hit and Run Tactics **142**
 Retreating and Returning 142
 Extinguishment and Follow-Up 143
Firing Operations **143**
 Burning Out 143
 Backfiring 143
  When to Burn Out or Backfire *143*
  Timing and Coordination *144*
  Control Lines for Firing Operations *144*
 Firing and Holding 145
Follow-Up **146**

    Before Leaving the Area 146
    Patrol Duties 147
Public Relations 147
    Dealing with the Media 148
    Dealing with the Public 149
Key Terms 149
References 151

Chapter 6 Water Supplies and Support 155
Water Sources 155
    Piped Systems 155
    Rural Water Sources 157
Mobile Water Operations 158
    Safe Operation, Driving, Off-Road Use 158
    Portable Water Tanks, Types, Sizes, Proper Location 159
    Water Shuttle Operations 159
    Nurse Tender Operations 160
    Pumping and Dumping Capabilities 160
    Tactical Tender/Tanker Operations 161

Chapter 7 Tactical Resource Support and Operations 165
Fire Crews 165
    Interagency Hotshot Crews 166
    Hand Crew Duties and Responsibilities 166
Heavy Equipment 167
    Dozers 168
    Tractor-Plows 168
    Road Graders and Other Mechanized Equipment 169
Hand Tools 170
    Cutting Tools 170
    Scraping Tools 171
    Fire Swatters (Flails) 172
    Wire Brooms 173
    Backpack Pumps 173
Fire and Aviation Management 174
    Helicopter Operations Safety Overview 174
    Helicopters 175
    Fixed-Wing Aircraft 176
    Smokejumpers 176
    Single-Engine Air Tankers (SEAT) 177
References 178

Chapter 8 Fire Control Tactics 181
Backfiring and Burning Out 181
    Backfiring 182
    Burning Out 183
Firing Devices 184
    Drip Torch 184

    Fusee 185
    Very Pistol/FireQuick 186
    Pneumatic Torch 187
    Propane Torch 187
    Terra Torch 187
    Plastic Sphere Dispenser (PSD) 188
    Helitorch 188

References 189

Glossary 193

Index 197

About this Manual

Changes in meteorological conditions over the past decade or more have created a greater risk for large wildland fires across North America. The western part of the U.S., in particular California, have seen an explosion of large fires in the hilly and mountainous parts of its jurisdictions. There are many training and informational sources for addressing wildland fires from agencies such as the U.S. Forest Service (USFA), the National Wildfire Coordinating Group (NWCG), and the U.S. Department of Agriculture (USDA).

These educational sources provide information for all responders across the country. However, there is a perception by many personnel and agencies, particularly in the central and eastern portions of the country, that this information is only for big fires that occur in California. They find it difficult to envision how this information applies to the fast-moving, flat terrain fires that they face on a daily basis. In reality, this information is applicable to these types of incidents, but the trick is to determine how it can be applied.

It is the intent of this manual to synthesize the information from these trusted sources and make them more applicable for personnel and agencies that face the flatter, fast-moving fires that are encountered in many jurisdictions. These types of fires are more commonly referred to as ground cover fires, as opposed to forest fires. The information contained in the manual will aid firefighters in attacking these fires in an organized, safe, and effective manner.

About the Author

Tom Richter worked his first wildland fire in June 1977 on the Black Hills National Forest. Since that time, Tom has served the wildland community through the U.S. Forest Service, the Wyoming State Forestry Division, and as a member of the Illinois Department of Natural Resources Interagency Wildland Fire Program. Tom currently serves the firefighters of Illinois as the Wildland and Prescribed Fire Program Manager at the Illinois Fire Service Institute at the University of Illinois, Champaign-Urbana campus. Tom also has the privilege of serving as the Deputy Incident Commander for the State of Illinois All Hazards Incident Management Team. Tom resides in Rochelle, Illinois, with his wife Ellen and serves the community there as Coordinator for the Ogle County Emergency Management Agency.

Introduction
A Common-Sense Approach to Ground Cover Fire Fighting

The reality of today's changing environment, political climate, and the introduction of technology into every facet of our lives is cause for us to honestly evaluate our approach and attitude toward all types of fire responses, including ground cover fires. Fire departments and agencies tasked with fire suppression responsibilities are being charged with protecting more lives with an ever-decreasing natural resource base and an ever-increasing realm of man-made, physical improvements than ever before in history. We can no longer afford, nor should we tolerate, the continued pace of firefighter deaths in this country. It is devastating in terms of morale, family losses, money, and the reflection on the way we conduct business.

One sure way to get on a path of best practices and to stay there, both to protect ourselves and the citizens we serve, is to "walk the walk" and implement the tools and knowledge passed to us so graciously by those who came before us and often learned through lessons and experiences that, in some cases, cost them their lives. We can no longer afford to be moths to the flames! We need to slow down, be methodical in our assessments and size-ups, implement the Incident Command System, and take time to ensure good accountability and effective resource management. It will save both lives and money. Believe it or not, by doing this we will be extensively more productive in our operations. It is the purpose of this manual to provide the information that you need to take a more reasonable, safer approach to deal with the types of ground cover fires that many types of fire departments commonly respond to today.

Fire Knows No Jurisdiction: Implementing the 3 Cs

Mutual aid is a vital tool for today's fire service. No single agency can "go it alone" when facing a major ground cover fire or interface disaster. As local agencies are faced with decreasing resources, these agencies must turn to their neighbors for assistance with increased fire protection. In recent years, we have witnessed a greater number of situations where neighbor helps neighbor, ground cover firefighters help structural firefighters, structural engines work on wildfires, and everybody is working together to complete the mission. As is to be expected, there are issues that need to be resolved whenever two or more agencies are working together. The keys to resolving these issues are the 3 Cs:

- Communication
- Coordination
- Cooperation

You will see these points emphasized throughout this manual. We must throw off the chains of ego, poor attitude, and petty turf wars and work towards recognizing experience and training that will build a future cadre of experienced firefighters and managers that can work shoulder to shoulder on an equal footing.

Communication

Agencies likely to work together must train together to:

- Gain valuable exposure to each other's capabilities.
- Expose equipment and safety limitations.
- Reduce or eliminate agency differences.
- Overcome the lack of training and experience in various areas of fire fighting.
- Develop a network for finding out about new tools.
- Coordinate and share responsibility.

Before the fire occurs, coordinated efforts between government, homeowners, fire fighting agencies, and firefighters are essential to ensure firefighter and civilian safety during fire fighting events. After the fire starts, fire fighting efforts must be performed within the context of standard operating procedures and best practices that mitigate risk to those performing their duties in the interface fire environment. These procedures and practices are covered in this manual.

Cooperation: Unified Command

When establishing an Incident Command where multijurisdiction has brought both ground cover and structural fire fighting forces together, establish a Unified Command where all organizations with jurisdictional responsibility are represented at the Command level. This ensures that an agency will not lose statutory or functional control of its respective resources; cost share can be implemented; facilities and support functions can be collocated; and everyone's concerns are addressed through a common set of objectives and a single incident action plan.

Honesty and the Risk Benefit Analysis

How many of you would knowingly sacrifice one of your own? The answer is simple — none of you. Yet, we commit folks to fields completely fenced in with running fire and light, flashy fuels. We knowingly place personnel and apparatus on ridgelines with unburned fuel between them and the fire. We knowingly, or without really assessing conditions, commit apparatus to fields that are thawing or have just received rain or into bottom lands that we know to be soft, just to have the vehicle become stuck with active fire in these same locations. We commit personnel to subdivisions with limited access and choked with ground cover fuels. To what benefit do we make these decisions? What are we really gaining? How many lives are we saving?

We emphasize in almost every training scenario that *we will risk a lot to save a lot, risk a little to save a little, or risk nothing to save what is lost.* How many of you practice this legitimate risk management process that has been documented and tested over time? How many of you discuss near-miss situations honestly and identify best practices to avoid similar situations? How many of you have refused an assigned task or offered an alternative solution to complete a mission after discovering conditions were unsafe and outside normal operating procedures instead of carrying on and luckily getting by unharmed? It is time to be honest with ourselves, those around us, and those under our command and ask ourselves: Is this really worth the potential loss of life? Will the risk taken benefit us in the grand scheme of the operation, or will the outcome be the same? The outcome is a foregone conclusion. Firefighters will eventually prevail. They will stop the fire, or it will run out of fuel or experience a change in conditions. The question will remain: At what cost and for what benefit?

Practice what we preach and teach, train hard, validate and implement best practices, truly assess the situations you find yourselves in, and weigh the cost. The information contained in this manual will help you make sound decisions when faced with ground cover fire incidents. Use this manual to guide your training and policy development, and then train your personnel effectively. Then, and only then, commit to a fight with a sound plan, the right resources, under the conditions and situations that will give support to the words AMERICA'S BRAVEST!

Chapter 1

Ground Cover Fire Behavior

Table of Contents

Types of Ground Cover Fires **4**
    Ground Fires 5
    Surface Fires 5
    Crown Fires 5

Ground Cover Fire Behavior **5**
    Ground Cover Fuels 5
    Weather 6
    Wind 6
        General Winds 7
        Local Winds 7
    Air Temperature 8
    Relative Humidity 8
    Precipitation 9
    Atmospheric Stability 9
    Topography 11
        Aspect 12
        Slope 13
        Terrain 13

Firelines **14**

Extreme Fire Behavior Characteristics **15**

Key Terms **16**

References **17**

Ground Cover Fire Behavior

Chapter 1

As a firefighter or supervisor, you should use the information in this chapter as a base for all your tactical decisions. This will help you understand the fire behavior you witness on arrival and what to expect in terms of **rate of spread**, direction, and fire intensity. This chapter is probably the most important part of the book, followed closely by Chapter 2, Firefighter Safety and Survival.

Ground cover fires can occur in vacant urban lots, parks, cemeteries, farm lands, and forests. Not only can they destroy hundreds of acres of natural resources, destroy houses and farm structures, and kill livestock and wild game, they place firefighters at risk. Ground cover fires include those in (**Figure 1.1**):

- Weeds
- Grass
- Field crops
- Brush
- Forests
- Similar vegetation

Figure 1.1 Examples of wildland fires in weeds and grass, brush, and a forest.

Because ground cover fires are unconfined, they differ from fires in burning buildings. The three main influences on ground cover fire behavior are fuel, weather, and topography. Weather is the most significant. Because local topography, fuel types, water availability, and predominant weather patterns vary from one region to another, the tools and techniques used to control ground cover fires also vary.

Once a ground cover fire starts, burning can be rapid. For your survival, you must learn how ground cover fires behave in a variety of conditions and how temperature and humidity, winds, fuel types, and terrain features influence **flame lengths** and rate of fire spread **(Figure 1.2)**.

Figure 1.2 A large-scale wildland fire as seen from an aircraft.

WARNING
Ground cover fires can be deadly to firefighters even if they are working in very light fuels or working during the overhaul phase of an operation.

Typically, two of the elements of the fire triangle — oxygen and fuel — are always present where ground cover is found. Adding a natural or human-made ignition source is the last remaining element needed to start a ground cover fire. Weather and topography also contribute to the intensity and spread of the fire.

Types of Ground Cover Fires

There are three basic types of ground cover fires based on the type and location of the fuel: ground fires, surface fires, and crown fires **(Figure 1.3)**.

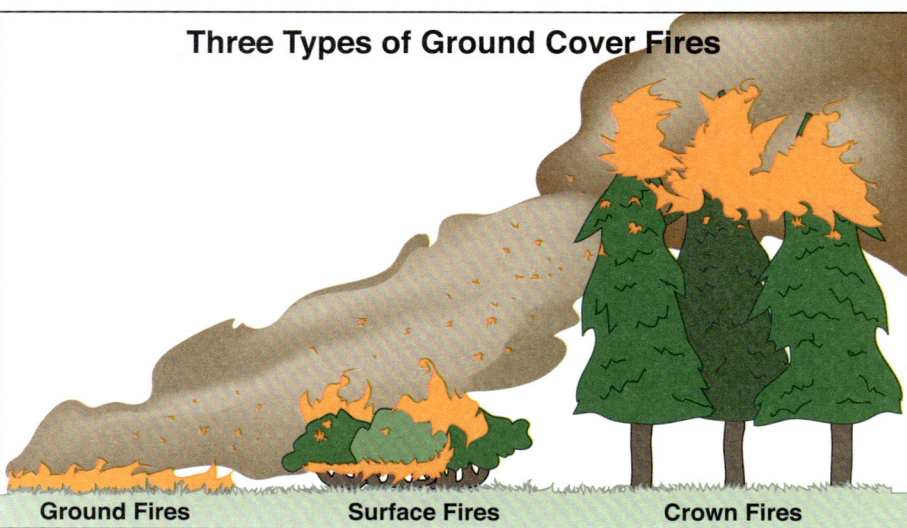

Figure 1.3 Examples of the three basic types of ground cover fires.

Ground Fires
Ground fires burn in the layer of dead organic matter (called humus) that generally covers the soil in forested areas. These are slow-moving, smoldering fires that can go undetected for months before they enter a flaming stage. Due to the composition of the fuel, these fires are generally limited to forests and are very difficult to extinguish.

Surface Fires
The surface or crawling fire is the most common type of ground cover fire, burning on the soil surface consuming low-lying grass, shrubs, and other vegetation. Surface fires can occur anywhere and can be natural or human caused.

Crown Fires
Crown or canopy fires are wind-driven, high-intensity fires that move through the tree tops of heavily forested areas. Ground or surface fires are sometimes the causes of crown fires. These extensions occur when the fire spreads upward through *ladder fuels* such as small trees, fallen timber, and vines to reach the forest canopy.

Ground Cover Fire Behavior
Typically, two of the elements of the fire triangle — oxygen and fuel — are always present where ground cover is found. Adding a natural or human-made ignition source is the last remaining element needed to start a ground cover fire. Weather and topography also contribute to the intensity and spread of the fire.

Ground Cover Fuels
Ground cover fuels are typically categorized based upon the location of the fuels as follows:

- **Subsurface or ground fuels** — Roots, peat, and other partially decomposed organic matter that lie just under the ground
- **Surface fuels** — Needles, duff, twigs, grass, field crops, brush up to 6 feet (2 m) in height, downed limbs, logging slash, and small trees on or immediately adjacent to the surface of the ground
- **Aerial fuels** — Suspended and upright fuels (brush over 6 feet [2 m]), leaves and needles on tree limbs, branches, hanging moss) physically separated from the ground's surface (and sometimes from each other) to the extent that air can circulate freely between them and the ground

  The following factors affect the burning characteristics of ground cover fuels:

- **Fuel size** — Small or light fuels burn faster than heavier ones.
- **Compactness** — Tightly compacted fuels burn slower than those that are loosely piled.
- **Continuity** — When fuels are close together, the fire spreads faster because of heat transfer. In patchy fuels (those growing in clumps), the rate of spread is less predictable than in continuous fuels.

- **Volume** — The amount of fuel present in a given area (its volume) influences the fire's intensity and the amount of water needed to achieve extinguishment.
- **Fuel moisture content** — Fuels that contain less moisture ignite more easily and burn with greater intensity (amount of heat produced) than those with a higher moisture content.

Weather

Some weather factors that influence ground cover fire behavior are the following:

- **Wind** — Fans the flames into greater intensity and supplies fresh air that speeds combustion and affects the rate and direction of fire spread; very large-sized fires create their own winds and/or weather systems **(Figure 1.4)**.

Figure 1.4 High winds pushing flames across a dirt road towards new fuel packages.

- **Temperature** — Closely related to relative humidity; primarily affects the fuels as a result of long-term drying.
- **Temperature differences** — Wind usually moves from areas of high pressure and heat to areas of low pressure and cold. Large wildfires can sometimes create their own wind systems due to the increases in temperature difference.
- **Relative humidity** — Significantly affects dead fuels that only gain moisture from surrounding air rather than their root system.
- **Precipitation** — Largely determines the moisture content of live fuels. Dead fuels (those easily ignited) may dry quickly; large, dead fuels retain this moisture longer and burn slower.

Wind

Wind is the horizontal movement of air relative to the surface of the earth. Wind is the most critical weather element affecting wildland fire behavior; it is difficult to predict and variable in both time and location.

Wind direction is the direction from which the wind is blowing (for example, a north wind means the wind is blowing from the north towards the south). This variability (especially in rough terrain) can pose safety and fire control problems that can result in firefighter injuries and fatalities. All firefighters must constantly monitor wind direction and wind speed. Wind impacts the fire environment by:

- Increasing the supply of oxygen to the fire
- Determining the direction of fire spread
- Increasing the drying of the fuels
- Carrying sparks and firebrands ahead of the main fire causing new spot fires
- Bending flames results in the preheating of fuels ahead of the fire
- Influencing the amount of fuel consumed by affecting the residence time of the flaming front of the fire

Stronger winds produce shorter residence time and therefore less fuel is consumed. There are two types of winds — general winds and local winds.

General Winds
Large-scale, upper-level winds caused by high and low pressure systems. If strong enough, general winds can influence ground cover fire behavior, but are generally modified in the lower atmosphere by terrain.

Local Winds
Found at lower levels of the atmosphere. Local winds are induced by small-scale (local) differences in air temperature and pressure and are best developed when skies are clear and general winds are weak. Terrain also has a very strong influence on local winds; the more varied the terrain, the greater the influence it has on local winds. Local winds can be as important to ground cover fire behavior as the winds produced by the large-scale pressure patterns. In many areas, especially in rough terrain or near large bodies of water, local winds can be the prevailing daily winds.

The types of local winds include:

- **Slope winds** — Local winds that develop in hilly and mountainous terrain where the differences in heating and cooling occur. During the day, the typical local wind pattern is upslope and downslope during the night. However, there will be cases where this rule does not apply.
- **Upslope winds** — Develop as air in the valleys, draws, and hillsides becomes warmer than the air at the top of the slope and begins to rise. The greatest upslope flow generally is about midafternoon; with speeds generally between 3 and 8 mph and can be gusty.
- **Downslope winds** — Generally occur after midnight and speeds range from 2 to 5 mph. The change from downslope to upslope can alter fire behavior from inactive to active in a matter of minutes.

NOTE: As firefighters have continually learned from others' injuries and fatalities, you must respect the dynamics of upslope flows knowing that fire runs surprising fast upslope. This regularly occurring behavior is one of the common denominators of fire behavior on tragedy fires.

- **Valley winds** — Produced by local temperature and pressure differences within the valley or between a valley and a nearby plain. Exceptions are valley winds that flow-up valley during the day and down-valley at night. As air in the valley warms, temperature and pressure differences within the valley or valley to adjacent plains results in an up-valley wind flow. The greatest up-valley winds occur mid to late afternoon. Up-valley wind speeds typically range between 10 and 15 mph (16 km/h and 24 km/h). Because of the large amount of air heated in the valley, up-valley winds develop after the upslope winds. Up-valley winds typically continue after sunset.

The air in the valley cools as the valley loses solar heating. The cool air drains down-valley, resulting in the down-valley wind. The greatest down-valley winds occur after midnight. Down-valley wind speeds typically range between 5 and 10 mph.

Air Temperature

Temperature is the degree of hotness or coldness of the air. Although we examine temperature every day, firefighters often pay no attention to the effects of this simple concept on fuels, firefighter fatigue, and the movement of air on a fire. When large areas of asphalt or concrete paving are heated by the sun, fires are influenced by creating voids in air space as mass amounts of warm air rises off these surfaces, creating a void. This void is filled by surrounding air thus pulling or pushing adjacent air masses, which in turn may influence fire behavior adjacent to these man-made improvements. The same holds true when open fields, sparse or areas devoid of fuels, heat up and cause the same effect near wooded areas on fire. There may be as much as a 60° difference between fuels in an open field and fuels under a canopy. The following are causes of changes in air temperature near the surface of the earth:

- Changing seasons. Seasonal and diurnal temperature changes can be large or small, depending on latitude, elevation, and topography.
- Alternations of night and day.
- Migrating weather systems. Proximity to the moderating influences of nearby oceans or lakes. Abrupt changes in temperatures can occur when migrating weather systems transport colder or warmer air into a region.

Heating of the earth's surface and the atmosphere is primarily a result of solar radiation from the sun; however, on a smaller scale, heat may be caused by a large fire. In the ground cover fire environment, direct sunlight and hot temperatures can preheat fuels and bring them closer to their ignition point, whereas cooler temperatures have the opposite effect. Above average temperatures are common on large fires. Many firefighter fatalities have occurred on fires where record high temperatures were set. Temperature is measured with a thermometer calibrated either to the Fahrenheit scale or the Celsius scale.

Relative Humidity

Relative humidity is the amount of moisture in the air usually expressed in percent. Relative humidity (RH) can range from 1% (very dry) to 100% (very moist). Low relative humidity is an indicator of high fire danger. Moisture in the atmosphere, whether in the form of water vapor, cloud droplets, or precipitation, is the primary weather element that affects fuel moisture content and the resulting flammability of ground cover fuels. The amount of moisture that fuels can absorb from or release to the air depends largely on relative humidity. Light fuels, such as grass and grain, lose moisture quickly with changes in relative humidity. Conversely, heavy fuels respond to humidity changes much more slowly.

Relative humidity values for extreme ground cover fire behavior vary over time and location and are different for different fuel types. As a general guideline, however, RH levels below 25% are an indicator of a problem or extreme fire behavior. Fuels in the southeast part of the United States and Alaska typically burn with considerably higher relative humidity than fuels in the western U.S.

Temperature and relative humidity have an inverse relationship. When temperature *increases*, relative humidity *decreases*. When temperature *decreases*, relative humidity *increases*. In the early morning hours, temperature typically reaches its lowest point and relative humidity reaches its highest point. As the sun rises and the temperature increases, relative humidity decreases. When the temperature reaches its maximum for the day (usually mid to late afternoon), relative humidity decreases to a minimum. This is when fine fuel moisture reaches its minimum, and we begin to see an increase in rate of spread and increased fire intensity.

Precipitation

Precipitation is liquid or solid water particles that originate in the atmosphere and become large enough to fall to the earth's surface. Fuel moisture is affected by the amount and also the duration of the precipitation. *Fine fuels react quite rapidly by precipitation since they gain or lose moisture usually within one hour.* Heavy fuels are not affected as drastically since they gain or lose moisture more slowly. A large amount of precipitation in a short time will not raise the fuel moisture as much as less rainfall over a longer period of time where the fuels can absorb more moisture before it runs off.

Atmospheric Stability

The degree to which vertical motion in the atmosphere is enhanced or suppressed is atmospheric stability. Stability is directly related to the temperature distribution of the atmosphere. Ground cover fires are greatly affected by atmospheric motion and the properties of the atmosphere that affect its motion. Many, if not all, of the great blowups and fatality fires have in some form been affected by unstable air masses. Most of the large land loss and fires with rapid rates of spread have been a combination of large convective lifts (upward air motion) associated with low pressure systems typical of unstable air days, low fuel moistures, and high temperatures. Surface winds, temperature, and relative humidity are most commonly considered and easy to measure in the fire environment. Less obvious, but equally important, is atmospheric stability and related vertical air movements that influence wildfire.

A *stable atmosphere* is defined as an atmosphere that resists upward motion. In the fire environment, the following visual indicators can give clues about the stability of the atmosphere:

- Clouds in layers
- Stratus type clouds
- Smoke column drifts apart after limited rise
- Poor visibility due to smoke or haze
- Fog layers and steady winds

The ability for firefighters to learn to tune into these visual indicators can often be the difference between a quick knockdown or a very long event.

The usual temperature structure of the lower atmosphere is characterized by a decrease in temperature with altitude. However, a layer where temperature increases with altitude (warm air over cold air) may exist. This layer is referred to as an *inversion*.

Under an inversion, fuel moisture content is usually higher, therefore decreasing fire spread rates and intensities. Updrafts containing smoke and warm gases generated by a fire are typically weak and will only rise until their temperature equals that of the surrounding air. Once this occurs, the smoke flattens out and spreads horizontally.

Increased ground cover fire behavior is almost certain when inversions break or lift as a result of heating the lower atmosphere by the sun or a fire. When an inversion breaks, watch for the following indicators:

- Increase in temperature
- Decrease in relative humidity
- Increase and/or shift in wind

Firefighters may encounter four (4) types of inversions: nighttime (often called **radiation**), subsidence, frontal, and marine. Nighttime and subsidence inversions are most common in the ground cover fire environment, although all inversions are important.

Nighttime (radiation). Inversions develop on calm, clear nights when radiational cooling of the earth's surface is greatest and can differ in strength depending on time of year. Nighttime inversions are easy to identify because they trap smoke and gases resulting in poor visibilities in valleys or drainages.

Subsidence inversion. The large-scale sinking of air associated with high-pressure systems cause subsidence inversions. As air from higher elevations in high-pressure systems descends to lower elevations, it warms and dries. The warming and drying of air sinking is so pronounced that saturated air (air with 100% RH) can produce relative humidity less than 5 percent in a very short period of time. If a high-pressure system persists for a period of days, the subsidence inversion may reach the surface with only very little external modification or addition of moisture. Skies are typically clear or cloudless under these high-pressure systems. Extended periods of above average temperatures and below-average relative humidity can dry out fuels to the point that burning conditions become severe. Firefighters must know and understand these conditions to ensure their safety. By constant monitoring of the weather and obtaining a spot weather forecast, firefighters can protect themselves from a severe situation and possible harm.

An *unstable atmosphere* is defined as an atmosphere that encourages upward motion. When the atmosphere is unstable, vertical motions increase and contribute to increased fire activity by:

- Allowing convection columns to reach greater heights
- Producing stronger indrafts
- Increasing the lofting of firebrands by updrafts
- Increasing the occurrence of dust devils and fire whirls Increasing the potential for gusty surface winds

Fire whirls are totally unpredictable and characterize an unstable and intensifying fire environment and are a key visual indicator of dangerous and problematic fire behavior. A fire whirl is a spinning vortex column of ascending hot air and gases rising from a fire and carrying aloft smoke, debris, and flame. Fire whirls range in size from less than one foot to over 500 (150 m) feet in diameter. Large fire whirls have the intensity of a small tornado (**Figure 1.5**).

Figure 1.5 Fire whirls can be extremely dangerous.

Your safety relies on knowing and recognizing an unstable atmosphere and having the ability to anticipate a change in fire behavior. Ground cover fires burn hotter and with more intensity when the air is unstable. Cold air over warm air represents an unstable condition.

Topography

Topography refers to the earth's surface features. The steepness of a slope affects the rate and direction of a ground cover fire's spread. Fires will usually spread faster uphill than downhill, and the steeper the slope, the faster the fire spreads **(Figure 1.6)**. Other topographical factors influencing ground cover fire behavior include the following:

Figure 1.6 Topography can influence the speed at which a ground cover fire spreads.

- **Aspect** — The compass direction a slope faces (aspect) determines the effects of solar heating. In North America, full southern exposures receive more of the sun's direct rays and therefore more heat. Ground cover fires typically burn faster on southern exposures.
- **Local terrain features** — Features, such as canyons, ridges, ravines, and even large rock outcroppings, may alter air flow and cause turbulence or eddies, resulting in erratic fire behavior.
- **Drainages (or other areas with wind-flow restrictions)** — These steep ravines are terrain features that create turbulent updrafts causing a chimney effect. Wind movement can be critical in *chutes* (narrow V-shaped ravines) and saddles (depression between two adjacent hilltops). Fires in these areas can spread at an extremely fast rate, even in the absence of winds and are potentially very dangerous.

Aspect

Aspect is the direction a slope is facing (its exposure in relation of the sun) **(Figure 1.7)**. The aspect of a slope generally determines the amount of heating it gets from the sun; therefore, determines the amount, condition, and type of fuels present.

South and southwest slopes are normally more exposed to sunlight and generally have:

- Lighter and sparser fuels
- Higher temperatures
- Lower humidity
- Lower fuel moisture

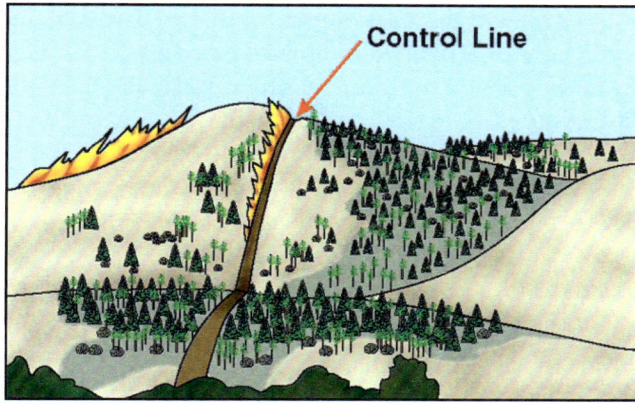

Figure 1.7 Aspect affects fire behavior in several ways.

They are the most critical in terms of start and spread of ground cover fires.

North and east facing slopes have more shade which causes:

- Heavier fuels
- Lower temperatures
- Higher humidity
- Higher fuel moistures

Slope

Slope is the amount or degree of incline of a hillside (a steep slope). Fires burn more rapidly uphill than downhill, which is a common denominator of fire behavior on tragedy fires. The steeper the slope, the faster the fire burns. This is because the fuels above the fire are brought into closer contact with the upward moving flames.

Convection and radiant heat help the fuel catch fire more easily. Another concern about steep slopes is the possibility of burning material rolling down the hill and igniting fuel below the main fire. The position of the fire in relation to the topography is a major factor in the resulting fire behavior. A fire on level ground is primarily influenced by fuels and wind. A fire which starts near the bottom of a slope during normal upslope daytime wind will normally spread faster and has more area to spread upslope than a fire that starts near the top of the slope.

Terrain

Certain topographic features can influence the wind speed and direction for small areas, independent of general weather conditions for an area. The shape of the country can also influence the direction of fire spread, rate of spread, and the intensity. When influenced by terrain, wind and fire act just like water in a stream. For example, fire will spill over and around obstacles and eddy on the lee side; fire will speed up in tight places, such as spaces between homes, tree lines, and streets; wind has the greatest influence on fire, more than any other component of the fire environment. Knowing, studying, and understanding the environment in your jurisdiction will help you anticipate fire behavior, making you safer and a stronger more proactive tactician. The following are definitions of the most common topographical features:

- **Box canyons**. Fires starting near the base of box canyons and narrow canyons may react similar to a fire in a wood burning stove or fireplace. Air will be drawn in from the canyon bottom creating very strong upslope drafts. These upslope drafts create rapid fire spread up the canyon, also referred to as the *chimney effect*. This effect can result in extreme fire behavior and can be very dangerous. Poor access and egress are generally the case with these features, one way in, one way out!
- **Narrow canyons**. Fire in a steep, narrow canyon can easily spread to fuels on the opposite side by radiation and **spotting**. Wind eddies and strong upslope air movement may be expected at sharp bends in canyon. Narrow canyons have poor means of access and egress with narrow roads and poor visibility. Most times there is no place to maneuver apparatus.
- **Wide canyons**. Alter prevailing wind direction by the direction of the canyon. Cross-canyon spotting of fires is not common except in high winds. Strong differences in fire behavior will occur on north and south aspects.
- **Ridges**. Fire burning along lateral ridges may change direction when they reach a point where the ridge drops off into a canyon. The flow of air in the canyon causes this change of direction. Ridge lines with unburned fuel between you and the fire is a setup for possible disaster. The confluence of any intersecting land features is a dangerous place to be.
- **Saddle**. Wind blowing through a saddle or pass in a rolling country or mountain range can increase in speed as it passes through the constricted area and spreads out on the downwind side with possible eddy action on the lee side of the saddle.
- **Barriers**. Any obstruction to the spread of fire, typically an area or strip lacking any flammable fuel. Barriers to fire include many things, both natural and man-made.

Natural barriers:

— Rivers and streams

— Lakes, ponds, and low-lying wetlands

— Rocky areas and slides

Fuels that have a high moisture content do not burn as well as others in the same area.

Man-made barriers:

- Roads
- Highways
- Reservoirs
- Fireline constructed by fire resources

Firelines

Firelines are parts of control lines along which fuel and sometimes earth is removed to create a fire stop **(Figure 1.8)**. Starting from an anchor point, a fireline is constructed some distance from the fire's edge and the unburned intervening fuel is allowed to self-extinguish. Creating the line involves using shovels, axes,

Figure 1.8 Firefighters throw dirt on a creeping ground cover fire as they create a fireline.

power saws, and other hand tools to remove all surface and subsurface fuels, such as roots, along the assigned fireline. To make a line as effective as possible, the following must be done:

- Remove all vegetation and debris from the line.
- Clear the line down to mineral soil.
- Widen the line as directed in order to provide a sufficient fire break depending upon height of the vegetation.
- Throw all burned/charred material into the black.
- Scatter all cut and unburned fuels into the green.
- Remove all branches that hang over the line.

Extreme Fire Behavior Characteristics

In recent times, it seems that all responders' responses to ground cover fires have been characterized by greater threats to responders and the public alike. Without almost any fires taking place in a true wilderness area, the threat to natural resources has been trumped by the need to protect public and private property. All firefighters take ground cover fires seriously; however, at times they often miss the signs and characteristics which give notice that this fire may be different. Like most of the concepts shared in this chapter, take note of the telltale things that will give clues to help make smart decisions and stay safer. This behavior is characterized by rapid rates of spread, frequent spot fires, trees **torching**, and flame lengths greater than 8 feet (2.4 m). Be aware of the following characteristics that give warning to impending problems and extreme fire behavior.

- Continuous fine fuels
- Heavy loading of dead and down fuels
- Plentiful ladder fuels
- Tight crown spacing
- Numerous firebrand sources
- Numerous snags
- High dead-to-live ratio of fuels
- Low relative humidity (less than 25%)
- Drought conditions
- Seasonal drying
- High temperatures (above 85° F [29° C])
- Steep slopes (greater than 30%)
- Chutes and box canyons
- Surface winds above 10 mph (15 km/h)
- Cumulonimbus development
- Dust devils and fire whirls
- Trees torching
- Well-developed smoke columns

Changes, such as fuel types, weather, topography, and fire behavior, often translate into safety hazards for the firefighters charged with controlling and extinguishing the fire. The volume and types of fuels available to a wildland fire can significantly influence a fire's behavior. Weather changes can increase or decrease fuel moisture, which can increase or decrease the susceptibility of the available fuels to ignition. The topography in which a wildland fire is burning can also influence the fire's behavior. Fires generally burn faster upslope than down.

Firefighters must understand the potential influences on fire behavior from these various phenomena for their personal safety and they must:

- Know how various fuels in the area may influence fire behavior.
- Be able to assess how weather can influence the direction and rate of fire spread.
- Know and understand how the terrain in which a fire is burning can influence the fire's behavior.

Key Terms

Fire Whirl — Spinning vortex column of ascending hot air and gases rising from a fire and carrying aloft smoke, debris, and flame. Fire whirls range in size from less than one foot to over 500 feet in diameter. Large fire whirls have the intensity of a small tornado (National Wildfire Coordinating Group (NWCG) *Glossary of Wildland Fire Terminology*).

Flame Length — The distance between the flame tip and the midpoint of the flame depth at the base of the flame (generally the ground surface), an indicator of fire intensity (National Wildfire Coordinating Group (NWCG) *Glossary of Wildland Fire Terminology*).

Radiation — Transfer of heat in straight lines through a gas or vacuum other than by heating of the intervening space.

Rate of Spread — The relative activity of a fire in extending its horizontal dimensions. It is expressed as rate of increase of the total perimeter of the fire, as rate of forward spread of the fire front, or as rate of increase in area, depending on the intended use of the information. Usually it is expressed in chains or acres per hour for a specific period in the fire's history.

Spotting — Behavior of a fire producing sparks or embers that are carried by the wind and which start new fires beyond the zone of direct ignition by the main fire (National Wildfire Coordinating Group (NWCG) *Glossary of Wildland Fire Terminology*).

Torching — Burning of the foliage of a single tree or group of trees or shrubs from the bottom up. Torching is a visual indicator of an intensifying surface fuels fire and changing fire environment associated with a destabilizing air mass (National Wildfire Coordinating Group (NWCG) *Glossary of Wildland Fire Terminology*).

References

National Wildfire Coordinating Group (NWCG) *Glossary of Wildland Fire Terminology*. Accessed online. https://www.nwcg.gov/glossary/a-z

Chapter 2

Firefighter Safety and Survival

Table of Contents

Operational Leadership 21
Risk Management 22
Standard Firefighting Orders 23
Turn Down (Refusing Risk) 26
Eighteen Watchout Situations 27
    Lookouts, Communications, Escape Routes, and Safety Zones (LCES) 31
        Lookouts 31
        Communications 32
        Escape Routes 32
        Safety Zones 32
Personal Protective Equipment 35
    Structural Gear 35
    Wildland Gear 36
        Wearing PPE on the Fireline 36
        Fire Shelter Basics 37
        How the Fire Shelter Works 38
        Determining Fire Shelter Location 39
    Agency Policy 40
Rules of Engagement 40
    Situational Awareness 40
    Look Up, Look Down, Look Around 41
    Look Up 41
    Look Down 42
    Look Around 42

    Avoiding Lightning 42
    30/30 Rule 43
Common Denominators of Fire Behavior on Tragedy Fires 43
Wildland Fire Apparatus Safety 44
    General Guidelines 44
    Off-Road Guidelines 44
        Capabilities and Limitations 45
        Hoseline Safety Guidelines 45
        Cautions in Off-Road Apparatus Operation ... 45
    Personnel Transport 46
Heavy Equipment Operations 47
    Heavy Equipment Hazards 48
    Seeing and Being Seen 48
    Heavy Equipment Operations Safety 48
        Hand Tool Safety 49
        Radio Communication 49
        Hand Signals 50
        Safety Tips for Firefighters 50
        Heavy Equipment and Underground Utilities ... 50
Hazardous Materials Situations 51
Operations Near Electrical Power Lines ... 53
Snag Safety 54
Key Terms 55
References 55

Firefighter Safety and Survival

In many ground cover fires, lives are not at risk until the first firefighters arrive at the scene. There is no grass, brush, timber, or structure worth a firefighter's life — only a reasonable chance to save another human life is worth that risk. You are personally responsible for your safety. Others, such as your supervisor, fire department, agency or contractor, are not responsible for your safety. To take responsibility for your safety, practice the following guidelines:

- Be in good physical condition.
- Maintain your mental health.
- Have a good attitude.
- Keep an excellent situational awareness while on duty and when engaged in fire fighting operations.
- Know and understand department/agency policy concerning ground cover fire operations.
- Be familiar with all the tools and equipment you may be asked to use.
- Know how to safely operate all the tools and equipment.
- Care for and wear ALL your personal protective equipment at all times, including during mop-up operations and while riding in the apparatus.

NOTE: Much of the following information is taken or adapted from the following resources: The National Interagency Fire Center (NIFC) https://www.nifc.gov/; Fire Shelters-The National Interagency Fire Center (NIFC) https://www.nifc.gov/fireShelt/fshelt_main.html; the National Wildfire Coordinating Group (NWCG) *Wildland Fire Incident Management Field Guide, PMS 210, April 2013*; and the National Wildfire Coordinating Group (NWCG) *Incident Response Pocket Guide*, January 2014, PMS 461.

Operational Leadership

Competent and confident leadership are the most essential elements of successful ground cover fire fighting (**Figure 2.1**). As a leader, the best way to show your leadership is to set the proper example concerning safety practices and ensure that personnel adhere to department, agency, or company policy with a

Figure 2.1 Operational leadership requires competent and confident leadership.

zero tolerance for violations concerning these policies. Operational leadership means providing purpose, direction, and motivation for firefighters working to accomplish difficult tasks under dangerous and stressful circumstances.

Use the following tasks to supervise other firefighters:

- Maintain accountability of assigned personnel, especially during incident operations.
- Enforce safe practices.
- Be a model for safe behavior.
- Assign fireline assignments only to those who are properly qualified and physically fit for the job.
- Analyze work situations at the beginning of each shift or new work assignment.
- Discuss safety procedures and hazard guidelines at the beginning of each shift (**Figure 2.2**).

Figure 2.2 The leader discusses safety procedures and hazard guidelines at the beginning of each shift.

- Respond immediately when an injury occurs. Be certain that medical treatment is provided in a timely manner.
- Observe work to be sure that it is done safely and efficiently.
- Monitor and enforce work/rest guidelines.
- Apply corrective action to eliminate accidents. Maintain a safe work attitude.
- Protect employees from retaliation for reporting unsafe conditions.

Risk Management

The foundation for all fire management activities is through risk management. Risk management is defined as the "process whereby management decisions are made and actions taken concerning the control of hazards and acceptance of remaining risk." You must identify, assess, and mitigate (or eliminate) the risks involved with any fire activity when possible and practicable. During the management decision of continuing or discontinuing the activity, everyone involved must consider the remaining risk acceptable and weigh it against the potential benefit. Risk management is practiced to minimize firefighters' exposure to inherent hazards in fire operations while still accomplishing management objectives.

Standard Firefighting Orders

The orders are guidelines that help firefighters identify and avoid high-risk situations. The *Standard Firefighting Orders* are organized in a deliberate and sequential way to be implemented systematically and applied to all fire situations. Every firefighter should know and follow them. A description of the orders follows:

1. Keep informed on fire weather conditions and forecasts.
2. Know what your fire is doing at all times.
3. Base all actions on current and expected fire behavior.
4. Identify escape routes and safety zones and make them known.
5. Post lookouts when there is possible danger.
6. Be alert, keep calm, think clearly, and act decisively.
7. Maintain prompt communication with your forces, supervisor and adjoining forces.
8. Give clear instructions and be sure they are understood.
9. Maintain control of your forces at all times.
10. Fight fire aggressively but provide for safety first.

> **1. Keep informed on fire weather conditions and forecasts.** Always request spot weather forecasts for fires that have the potential for active fire behavior, exceed initial attack, or are located in areas where Red Flag Warnings have been issued. In addition, consider requesting a spot weather forecast for nonfire incidents including hazmat or search and rescue activities.
>
> A spot weather request requires the following basic elements:
> - Name and type of incident
> - Location by latitude/longitude or by one-quarter section
> - Incident size
> - Elevation (at top and bottom of incident)
> - Fuel type
> - Sheltering (full, partial, unsheltered)
> - Fire character (ground fire, crowning, spotting, etc.)
>
> The following weather observations include:
> - Location on the fire
> - Elevation of observation

- Aspect of observation
- Time of observation
- Wind direction
- Wind speed
- Dry bulb
- Wet bulb
- Relative Humidity
- Dew point

2. Know what your fire is doing at all times.
- Observe the fire from a vantage point.
- Scout ahead.
- Send out reliable scouts (who then report back).
- Observe the fire from a helicopter or other aircraft if available.
- Inform all crew members of the current situation.
- Review the incident action plan (IAP).
- Monitor assigned radio frequency.

3. Base all actions on current and expected fire behavior. Base every action on what a fire is doing and what you think it might do. Firefighters should ask the following:
- What is the fire doing now? Can it be controlled with the resources at scene or en route? Do more resources need to be requested?
- What is the fire likely to do later? Can it be flanked? Should we back off for an indirect attack?
- What action is being taken now? Is equipment being used effectively? Are we making the attack at the right spot?
- What is the weather in the fire area? What is the weather likely to do?
- What type of fuel is burning? Can we build control lines fast enough? Is the fuel a type that can cause spot fires?
- What type of fuel is the fire heading toward? Do we have the proper tools? Will we need aircraft or mechanized equipment?

4. Identify escape routes and safety zones and make them known. Follow this order at all times. It is especially important when crews are traveling cross-country to a fire that has no control lines or to an area with which they are unfamiliar. The following are some good areas to select:
- *Safety zones* — Identify or construct them. Let their locations be known to all.
- *Burned area* — Make sure it is close enough to be reached.
- *Natural barriers* — Locate rocky areas, riverbeds, streams, lakes, and slide areas; let others know about them.
- *Escape route* — *Have one*, even if it must be constructed. Mark a safe route into the burned area, or cut one if the brush is too thick for travel.

Be aware of the following with escape routes:

- Be cautious of areas that are not completely burned out.
- Communicate to crew members where escape routes are and how to travel to them. Inform firefighters what to do when they get to a safety zone.
- Change the escape route as the fire progresses, perhaps more than once.
- Use extreme caution if the escape route passes through a chute, gully, or saddle.

5. **Post lookouts when there is possible danger**. Lookouts must be in constant communication with firefighters or supervisors. If radios are not available, you may have to use a system of visual signals. Also, enough lookouts must be used to maintain visual contact with both the fire and the crew. Some situations that warrant a lookout are as follows:

- When the head of the fire is not visible to the crew
- When felling snags
- When personnel and mechanized equipment are working closely together
- When falling rocks could strike someone, or burning material could cross the control line
- When an obvious hazard, such as a snag, cannot be felled
- When airdrops are being made nearby
- When firing operations are being conducted nearby

6. **Be Alert. Keep Calm. Think Clearly. Act Decisively**. When faced with a possible life-threatening problem, crew members and leaders must keep calm and analyze what is happening. Panic can endanger the leader as well as the crew. After evaluating the situation, they should decide how to deal with it, and ***do it***! Staying calm is not the same as moving slowly. Do whatever you must to save yourself and your crew, and do it quickly! It is better to flee five minutes too soon than five seconds too late.

Factors that can affect decision-making include:

- Fatigue. Stay in shape, eat right, and get your sleep — and while on the fireline, drink lots of fluids.
- Heat stress. Remember to drink. Do all the same things that you would do for fatigue!
- Smoke (carbon monoxide). Reduce your exposure to it.
- Stress. Know and understand the life-threatening situation you are in.

7. **Maintain prompt communication with your forces, supervisor, and adjoining forces**. Adequate communication is essential to fireline safety. Use the following guidelines:

- Maintain good communication within each unit, between the unit and any scouts or lookouts, and with other fire fighting units.
- Identify your adjoining forces.
- Communicate face-to-face or by radio, telephone, or any other reliable means.

8. Give clear instructions and be sure they are understood. The likelihood of accidents occurring is increased if vague or ambiguous oral instructions are given instead of concise, written instructions. Oral instructions are sometimes incomplete and are more likely to be misinterpreted or forgotten. Crew leaders must verify the following when given an assignment:

- What to do (objective)
- How to do it
- Where to go
- Where to finish
- When to finish
- With whom to coordinate
- Identity of the supervisor on the line
- Identity of the relieving person
- Expected duration of attack
- Available transportation to and from the fireline
- Other pertinent information such as emergency procedures and safety considerations

9. Maintain control of your forces at all times. Consider the capabilities and limitations of crew members when making work assignments. Crew members must:

- Be rested and ready to work.
- Inspect tools, coordinate work with available equipment, and make provisions for safety.
- Exhibit command presence and utilize the chain of command in the fireground organization.

10. Fight fire aggressively, but provide for safety first. Aggressive action is the key to a successful fire suppression operation; however, safety is the first priority. Crew leaders should:

- Analyze the situation.
- Make an attack consistent with accepted practices and methods under the existing conditions.

Turn Down (Refusing Risk)

A **turn down** (also called *refusal of risk*) is a situation where an individual firefighter or crew leader has decided that he or she cannot carry out an assignment as given *and* is unable to negotiate an alternative solution. Individuals have an obligation to identify safe alternatives for completing the assignment. Turning down an assignment is one possible outcome of managing risk. Turning down an assignment must be based on a reasonable assessment of risks — usually the *10 Standard Firefighting Orders* and/or the *18 Watchout Situations.*

The individuals directly inform their supervisor that they are turning down the assignment as given. Individuals may turn down an assignment as unsafe when:

- A violation of safe work practices exists.
- Environmental conditions make the work unsafe.
- They lack the necessary qualifications or experience.
- Defective equipment is being used.

To document the turn down of the assignment as given, the individual directly informs his or her supervisor. Notify the Safety Officer immediately upon being informed of the turn down. If there is not a Safety Officer, notify the appropriate Section Chief or the Incident Commander. This notification provides accountability for decisions and initiates communication of safety concerns within the incident organization. If the supervisor asks another resource to perform the assignment, inform the resource asked to perform the assignment that it was turned down and the reasons why it was turned down and the reasons for the turn down.

If an unresolved safety hazard exists or an unsafe act was committed, the "individual" should also document the turn down by submitting either a SAFENET (ground hazard) form or a SAFECOM (aviation hazard) form in a timely manner. These actions do not stop an operation from being carried out. This protocol is integral to the effective management of risk as it provides timely identification of hazards to the chain of command, raises risk awareness for both leaders and subordinates, and promotes accountability.

> **NOTE:** The SAFENET system provides a way for frontline firefighters and support staff to report unsafe or unhealthy situations and near misses. Fire managers can use SAFENET to hear about and correct hazardous conditions, collect, and track important safety data and identify trends that could be developing.

Eighteen Watchout Situations

Much of the risk of fire fighting can be reduced if firefighters are alerted to the *18 Watchout Situations*. Experience over many years and in countless ground cover fires have produced a list of fireline situations that resulted in firefighters being killed. To avoid finding themselves in the same danger, firefighters must be familiar with the following list of situations:

18 Watchout Situations

1. **Fire not scouted and sized up.**
2. **In country not seen in daylight.**
3. **Safety zones and escape routes not identified.**
4. **Unfamiliar with weather and local factors influencing fire behavior.**
5. **Uninformed on strategy, tactics, and hazards.**
6. **Instructions and assignments not clear.**
7. **No communication link with crew members/supervisors.**
8. **Constructing line without safe anchor point.**
9. **Building fireline downhill with fire below.**
10. **Attempting frontal assault on fire.**

> 11. Unburned fuel between you and the fire.
> 12. Cannot see main fire, not in contact with anyone who can.
> 13. On a hillside where rolling material can ignite fuel below.
> 14. Weather is getting hotter and drier.
> 15. Wind increases and/or changes direction.
> 16. Getting frequent spot fires across line.
> 17. Terrain and fuels make escape to safety zones difficult.
> 18. Taking a nap near the fireline.

The "Watchout" list identifies some specific situations where the fire-related risk is higher than normal. Any firefighter in such a situation must be especially alert for changes in fire behavior that might increase the danger. The listed items are expanded as follows:

1. **Fire not scouted and sized up.** Crews are in danger if they cannot see the entire perimeter to fires or parts of fires and have not had the opportunity to adequately perform a size-up. To provide a reasonable degree of safety, crews must be aware of where the fire is and what it is doing.

2. **In country not seen in daylight.** When crews are in an area that was not seen in daylight, they may not have the topographical information needed to work in an adequate degree of safety. The situation may be unsafe due to the shape of the land, the density of the vegetation, and the distances between points.

3. **Safety zones and escape routes not identified.** Crews or individual firefighters on the fireline who cannot locate safety zones and escape routes should stop what they are doing until they learn where these critical safety features are located. Obviously, judgment must be used about when and where to stop, but it should be done immediately.

4. **Unfamiliar with weather and local factors influencing fire behavior.** The crew may find the local microclimate or burning conditions different from those in other areas. Safely coping with different conditions may require a conscious change in strategy or tactics.

5. **Uninformed on strategy, tactics, and hazards.** Crews can be placed in serious jeopardy by not knowing what the plan of attack is and how they fit into it. They may find themselves in the path of danger such as airdrops or firing operations. Receiving assignments face-to-face is always best and helps provide clarity.

6. **Instructions and assignments not clear.** The results can be both unproductive and dangerous. Once a firefighter or a crew is on the fireline, it may be too late to get orders clarified.

7. **No communication link with crew members or supervisor.** This can result in critical, lifesaving information not being passed up or down the chain of command. Maintaining reliable communications between all levels is an absolute necessity.

8. **Constructing line without safe anchor point.** While this may have to be done under some extraordinary circumstances, it is a dangerous practice that risks the crew being flanked and possibly surrounded by a fire. The decision to take this risk must be made only by a very experienced firefighter who has carefully weighed the benefits against the risk and established LCES.

9. **Building fireline downhill with fire below.** It is very hazardous to build a line downhill (or make a hose lay in the green) toward a fire below. Fire normally burns faster uphill than downhill. There is a greater risk of the fire flanking the crew working downhill. Also, convected heat, smoke, and flame rising upslope make it difficult for firefighters to breathe or see clearly, and they are likely to have very poor footing.

10. **Attempting frontal assault on fire.** A crew attempting a frontal assault on a fire (from the green) is in a dangerous position, especially if it has too few hoselines or hoselines that are too small. The fire may overrun the firefighters or spot behind them.

11. **Unburned fuel between you and fire.** It is dangerous for firefighters to be in any type of ground cover with unburned fuel between them and a fire. They are actually in the green and are susceptible to being overrun by the fire as they attempt to move through the unburned fuel to reach the burned area. It is critical to have readily available escape routes and safety zones.

12. **Cannot see main fire, not in contact with someone who can.** If firefighters are working out of sight of the fire and they are not in contact with anyone who can see the fire, an unseen blowup can put them in danger of being overrun. Post one or more lookouts who can see the progress of the fire.

13. **On a hillside where rolling material can ignite fuel below.** If rolling materials start spot fires below, a new fire may run upslope toward the crew. Since fire can spread rapidly upslope, firefighters are not likely to be able to outrun the fire.

14. **Weather becoming hotter and drier.** When the weather becomes hotter and drier, fires become increasingly active. New smoke may appear within the burn, smoldering duff supports visible flame, and up-canyon winds may start to blow through the ravines and across control lines. Spot fires may increase in number, and smoldering spot fires may be fanned back into life. Pyrolysis increases, and fuels become more susceptible to ignition. The likelihood of crowning increases, even during mop-up.

15. **Wind increases and/or changes direction.** Be aware of winds increasing or changing direction. Wind flattens out the flames, which results in the ignition of new fuels and increases the rate of spread. Wind blowing from the green into the black may suddenly reverse direction and blow hot materials or flames into new fuels.

16. **Getting frequent spot fires across line.** If spot fires across the line are numerous, it may be difficult to reach them while they are still small. However, if they are not controlled, they can combine into area ignition. Spot fires can also develop into separate major fires, and crews can suddenly find themselves and their equipment between two or more fires.

17. **Terrain and fuels make escape to safety zones difficult.** A crew some distance from the burned area or another safety zone can be in terrain or cover that makes travel difficult and slow. Slopes present hazards such as falling rocks and the potential for slipping. Irregular terrain can put firefighters and other personnel out of sight of the fire. Heavy cover may also restrict their ability to see the fire and may obscure escape routes.

18. **Taking a nap near fireline.** A sleeping firefighter can be overrun by a fast-moving fire or by heavy equipment operating along the fireline. Crew members should be allowed to sleep on the fireline only in safety zones and only when lookouts have been posted. Drowsiness may also be an indication of carbon monoxide poisoning.

Look for the following hazards during fire operations:

- **Unstable trees** — Trees that have been weakened by age or fire and may collapse.
- **Animals** — Animals that have escaped the fire.
- **Insects** — Usually more of a nuisance than a hazard; some insect stings can be fatal to persons with allergies. Repellant sprays and first aid creams should be carried when needed.
- **Extreme temperatures** — Extreme hot temperatures can cause exhaustion, dehydration, heat exhaustion, and sunburns. Cold temperatures could lead to hypothermia.
- **Electrified fences** — Electrified fences have caused numerous firefighter deaths. All wire strand fences should be considered electrified until proven otherwise.
- **Electrical power lines** — Ground cover fires can cause power poles to fall and power lines to break. See the section on Class C fires for safety tips.
- **Explosives** — Explosives and unexploded ordnance may be found around military training areas, near construction sites, and in areas open to hunting. Do not touch or move any explosives that have been exposed to fire, and establish a perimeter around them.
- **Hazardous materials** — Broken pipelines, storage tanks, oil and gas wells, and storage buildings can create hazardous materials hazards when exposed to fire. Treat these situations like hazardous materials incidents, establishing a perimeter, and withdrawing a designated distance.
- **Rolling or falling debris** — In rough terrain, rocks, burned vegetation, and limbs can fall and strike you or create a slipping or tripping hazard.
- **Pits or shafts** — Loose debris may cover abandoned mine shafts, pits, and natural sink holes.
- **Animal traps** — Traps used for hunting may be hidden under brush. In areas where illegal activities have been reported, these devices may be used to protect them.

Lightning is a unique nonfire hazard that requires special precautions. The lightning that started a ground cover fire can also injure or kill firefighters. Take the following precautions to protect yourself from lightning hazards:

- Be aware of the weather.
- Do not stand under tall, isolated trees.

- Stay away from open water, metal objects, equipment, or wire fences.
- Find suitable shelter when lightning could affect your safety.

In a forested area, seek shelter in a low ravine. If you are in a flat field and feel your hair stand on end, it is an indication that lightning is about to strike. Drop to your knees and bend forward, putting your hands on your thighs. Do not lie flat on the ground.

Finally, most line-of-duty deaths at ground cover fires result from heart failure. At ground cover fires, monitor your stress level, check heart and lung rates, stay hydrated, and use the rehabilitation facilities that are provided.

Lookouts, Communications, Escape Routes, and Safety Zones (LCES)

LCES serves to remind crew leaders and supervisors of the essential elements involved in providing for the safety of their crews working on the line (**Figure 2.3**). LCES must be established and known to all firefighters before it is needed. When tactical plans are made, check them against LCES to make sure that these safety considerations are included. If safety plans are not included, do not implement the tactical plans until these safety provisions are incorporated.

Lookouts

A **lookout** is a firefighter who is assigned to continuously observe a fire and warn the crew when there is danger from the fire (**Figures 2.4 a and b**). A lookout must have experience, be competent, and trustworthy. It is important to post the lookouts to see both the crew and the fire when crews are assigned to work in drainages or other areas where they cannot see the fire, but the fire front is relatively close. Always have enough lookouts at good vantage points. Make available to your lookouts the crew locations, escape and safety locations, and trigger points. Also provide your lookouts with a map, weather kit, watch, and an IAP.

Critical Safety Considerations

Lookouts
Know where the fire is and where it is going.

Communications
Know who is operating above, below, and adjacent to you.

Escape Routes
Know more than one way out of the area you are working in.

Safety Zones
Know how to quickly get to an area of refuge.

Figure 2.3 Critical safety considerations for working on a ground cover fire.

Figures 2.4 a and b a) A lookout watches a wildland fire's progress. b) The lookout uses hand signals to alert the crew.

The lookout should be able to see both the fire and the firefighters and understand the fire behavior he or she sees. If there is a sudden or unexpected change in fire behavior that might threaten the crew, the lookout's function is to notify the crew in time for it to retreat to a safety zone or leave the threatened area. However, it is the responsibility of every crew member to watch for hazards, such as blowups, rolling rocks, and falling snags, and to warn the others.

Communications

It is imperative to have some form of reliable communication between the lookouts and the crew leader. The crew leader must sound the alarm early, not late. This is most often done by radio, and radio frequencies need to be confirmed. Radios are not always reliable in some areas such as mountainous terrain. Therefore, establish backup procedures and check-in times. If radios are not available or not effective, another form of communication must be used. Provide updates on any situation change. In these cases, direct voice communication is best, but using a system of hand signals, signal flags, or signal mirrors can work if the signals used are clearly understood. In more urbanized areas, cell phones and pagers may be used.

Escape Routes

An **escape route** is a path by which firefighters can rapidly leave an area of danger and find an area of safety. Use the following procedures when identifying an escape route:

- Have more than one escape route in advance of its need, and mark it for day or night.
- Time the routes considering the slowest person on the team, firefighter fatigue, and temperature factors.
- Evaluate escape time vs. rate of spread.
- Inform all members of the crew where the routes are located.
- Escape routes must be positively identified whenever firefighters are required to work in the green near the fireline, especially if there is unburned fuel between them and a fire.

According to the National Wildfire Coordinating Group (NWCG) Flagging Standards, use fluorescent pink flagging with the words ESCAPE ROUTE in black with black dots. The color lime-green should no longer be used to identify safety zone or escape routes. However, crews with colorblind individuals may wish to mark escapes routes and safety zones with both hot pink and lime green flagging.

Normally, escape routes should not be located above a fire burning uphill on a slope. Escape routes may lead into the black or farther into the green away from the fire such as to a safety zone. As a fire front progresses, the adequacy of previously identified escape routes changes also. New escape routes may have to be selected from time to time, and the location of these new routes marked and communicated to the crew.

Safety Zones

A **safety zone** is an area into which firefighters can survive without a fire shelter. Firefighters can escape if conditions deteriorate or fire behavior increases to a point that current tactical operations become ineffective or unsafe. When crews must work in heavy fuels, a safety zone should be maintained as close by as possible in case of a blowup. A safety zone must contain little or no combustible vegetation and be large enough for all crew members to deploy their fire shelters safely if necessary. However, survival in a safety zone should not depend upon deploying a fire shelter. If a safety zone is large enough and the fuels light enough, firefighters should not need to deploy their fire shelters within the zone.

To be effective, safety zones should have a radius equal to at least four times the flame length. Depending on the situation and the resources available, the larger the safety zone the better it is. Escape time and safety zone size requirements will change as fire behavior changes.

Natural and Constructed. Safety zones are open areas characterized by an absence of fuel available to the fire. They may be green meadows or naturally barren areas, such as rock slides or cliffs, or they may be streams or other bodies of water **(Figure 2.5)**. Safety zones may also be areas that previously burned out fully. In the absence of natural safety zones, they may be constructed by scraping away surface fuels down to mineral soil. In heavy fuels, mechanized equipment may be needed to construct safety zones, but in light fuels hand tools may be used. Safety zones can also be created by burning out. However, use caution because burning out can create additional safety problems such as drawing the fire to your location more rapidly.

Figure 2.5 A stream bed is a natural safety zone.

Substantial Structure. Any substantial structure may provide an area of refuge. The level of protection provided by a structure can be enhanced by burning out around it. Even if the structure ignites and is eventually lost to the fire, it can provide a refuge for firefighters long enough for the vegetation fire to burn past before they have to exit the structure.

Engine. If firefighters are about to be overrun by an intense and fast-moving ground cover fire, they can take refuge in the cab of their engine. However, newer engines may have a considerable amount of fiberglass or plastic used in body construction. These synthetic materials may ignite easily and force firefighters to abandon the engine before it is safe to do so. Firefighters should use the following guidelines:

- Position the engine in an area where the least amount of vegetation is available as fuel to the fire.
- Close all vents, and roll up the windows once in the cab.
- Cover the inside of the windows and windshield with one or more fire shelters **(Figure 2.6, p. 34)**.
- Use self-contained breathing apparatus (SCBA) when the smoke reaches the vehicle. SCBA also helps protect the occupants if the plastic components inside the cab begin to degrade and off-gas due to the heat **(Figure 2.7, p. 34)**.

Firefighter Safety and Survival • Chapter 2 33

Figure 2.6 Firefighters can take refuge in the cab of their engine if they are about to be overrun by an intense ground cover fire. Therefore, the inside of the windows and windshield should be covered with one or more fire shelters.

Figure 2.7 Firefighters should use SCBA when the smoke reaches the vehicle.

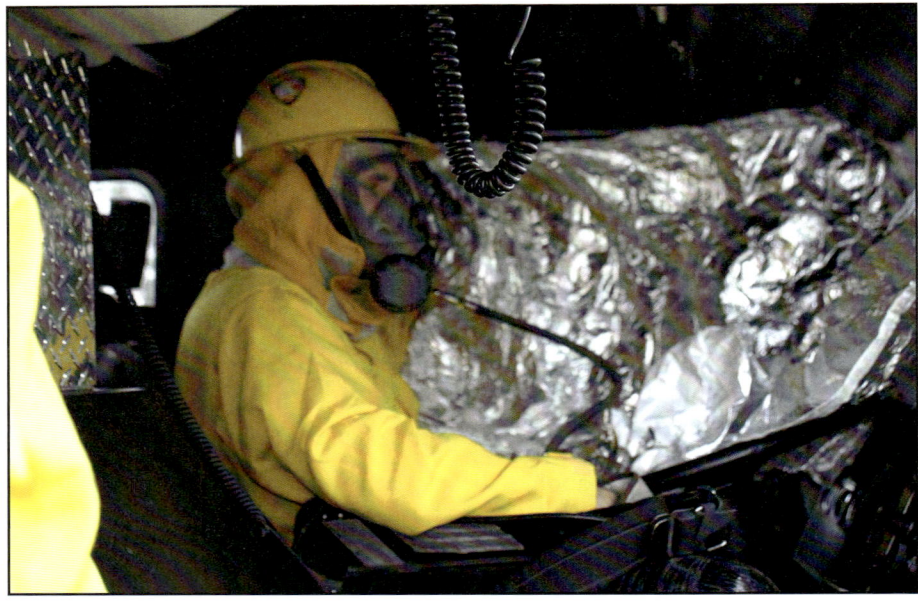

- Leave the engine in road gear with the motor running at a high idle to prevent stalling due to oxygen deficiency. This may allow a rapid escape from the area if the opportunity arises.
- *Do not* take a charged hoseline into the cab. This would provide an opening for smoke and hot embers to enter the cab, and the water could cause steam burns to the occupants.
- Partially discharge spare SCBA cylinders inside the cab to provide positive pressure and help prevent smoke from intruding.
- Do not lock engine cab doors as this may slow the escape of those inside or prevent a firefighter outside from getting into the cab.

- Crouch below the window line to reduce the effects of radiant heat.
- Do not park the engine above a chute or gully if possible.

> **WARNING**
> *Do not* take a charged hoseline into the cab. This provides an opening for smoke and hot embers to enter the cab, and the water could cause steam burns to the occupants.

Personal Protective Equipment

Firefighter safety is significantly enhanced through the use of proper personal protective equipment for the type of fire being fought. Personal protective equipment (PPE) for structural fire fighting and that for ground cover fire fighting is quite different, except for the fire-resistant materials of the protective clothing. When fighting a ground cover fire, firefighters should not wear structural gear. Contract workers must be provided with protective clothing if they are to be outside the incident base. Only personnel wearing proper personal protective clothing for the job they are assigned should be allowed near a fire (**Figure 2.8**). All personal protection items must meet applicable local, state/provincial, and national standards.

Structural Gear

Firefighters do not normally need their structural gear when protecting a structure from an approaching ground cover fire. Firefighter structural gear only becomes important when the ground cover fire ignites the structure and interior fire fighting becomes necessary or if wildland gear is unavailable.

In the past, specifications for personal protective equipment used in structural fire fighting were contained in several NFPA® standards. However, with the exception of self-contained breathing apparatus, specifications for all structural gear are now contained entirely in NFPA® 1971. This gear is designed to be functional in the activities associated with fighting structure fires, but it is generally too bulky, too hot, and too heavy to be practical for use in ground cover fires. However, wildland firefighters trained in interior structure fire fighting techniques should have structural turnout gear available to them when assigned to structure protection in the wildland/urban interface. Firefighters should wear additional safety equipment including approved chaps, gloves, hard hat, and eye and hearing protection.

Figure 2.8 Personnel must wear proper personal protective clothing for the job to which they are assigned.

Wildland Gear

NFPA® 1977, *Standard on Protective Clothing and Equipment for Wildland Fire Fighting* (2016), states the minimum requirements for the following protective wildland gear:

- Garments (shirt, pants)
- Helmets
- Work gloves
- Footwear
- Face/neck shrouds
- Goggles
- Chain saw protectors
- Driving gloves
- Load-carrying equipment

According to NFPA® 1977, all equipment used for wildland fire fighting must be labeled. The label must contain all pertinent information regarding that product and be attached to each article of personal protective equipment.

Wearing PPE on the Fireline

When assigned to the fireline and when flying in helicopters, all personal protective equipment (PPE) will meet or exceed agency policy **(Figure 2.9)**. Firefighters should wear the following protective clothing and equipment:

- Flame-resistant clothing. Do not wear flame-resistant clothing when the fabric is so worn as to reduce the protection capability of the garment, or it is so faded as to significantly reduce the desired visibility qualities.
- No clothing made of synthetic materials, even undergarments, which can burn and melt on the skin.
- Clean flame-resistant clothing. Clothing should be replaced whenever soiled, especially when stained with petroleum products.
- A minimum of 8-inch-high (200 mm), laced-type exterior work boots, with Vibram-type, melt-resistant soles **(Figure 2.10)**. The 8-inch (800 mm) height requirement is measured from the bottom of the heel to the top of the boot. Alaska is exempt from the Vibram-type sole requirement.
- Eye and face protection whenever there is a danger from material being thrown back into the face.
- Fire shelters (determine and comply with host agency requirements regarding fireline suppression assignments or follow agency's requirements if they are more restrictive.)
- Hard hat and leather gloves **(Figures 2.11 a and b)**.
- Hearing protection when working with high-noise-level fire fighting equipment, such as helicopters, air tankers, chain saws, pumps, etc.
- Approved chaps, gloves, hard hat, and eye and hearing protection when operating chain saws and sawyers. (A *swamper* is a worker who helps clear away brush, limbs, and small trees.) Swampers should wear chaps when the need is demonstrated by a risk analysis considering proximity of the chain saw to the sawyer, the slope, fuel type, etc.

Figure 2.9 All personal protective equipment will meet or exceed agency policy when assigned to the fireline.

Figure 2.10 Vibram-type, melt-resistant soles.

Fire Shelter Basics

Fire shelters are one of the most critical pieces of personal protective equipment that have saved many lives and prevented many serious burn injuries. The fire shelter is an aluminized cloth tent that offers protection in a fire entrapment situation by reflecting radiant heat and providing a volume of breathable air **(Figure 2.12)**. The shelter is comprised of two layers:

- Outer layer — Woven silica laminated to aluminum foil. The foil reflects radiant heat and the silica cloth slows the transfer of heat to the inside of the shelter.
- Inner layer — Fiberglass laminated to aluminum foil. The inner layer of foil prevents heat from being reradiated inside the shelter, and it prevents gases from entering the shelter.

When the two layers of materials are sewn together, the air gap between them provides additional insulation.

Fire shelters should be viewed only as a last resort when there is no way out of a life-threatening situation. Firefighters must also guard against becoming overconfident because they have a shelter and must not place themselves in danger unnecessarily. Fire shelters do not reduce the need for proper scouting, posting lookouts, and establishing escape routes and safety zones. All fireline personnel must carry fire shelters during all phases of fire suppression. Personnel must use the following guidelines:

- Inspect shelter according to manufacturer's instructions. Protect it from damage.
- Pack it where you can get it out quickly, even while running.
- Never keep it inside your pack where it would be hard to get to it in a hurry **(Figure 2.13)**.

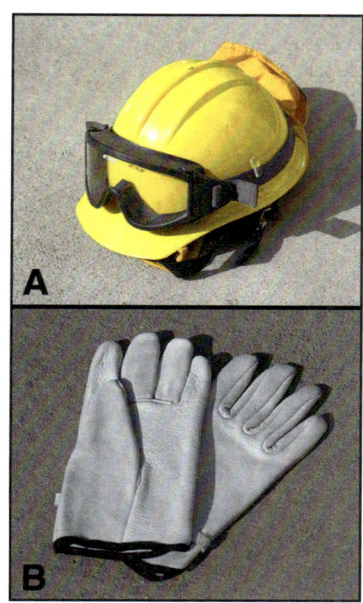

Figures 2.11 a and b Hard hats and leather gloves.

Figure 2.12 The fire shelter is an aluminized cloth tent that offers protection by reflecting radiant heat and providing a volume of breathable air.

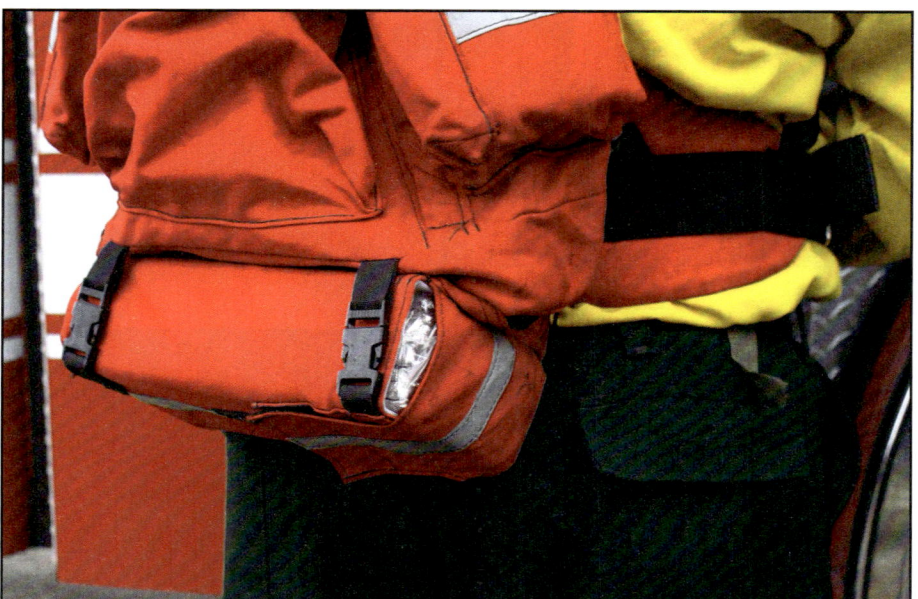

Figure 2.13 As shown in this photo, keep your fire shelter in a place where it is easily accessible.

- Prevent damage to your shelter by not sitting on it, using it as a pillow, or placing heavy objects on it.
- Keep shelter away from sharp objects that may puncture it.
- Do not place heavy objects on top of a shelter.
- Avoid rough handling of a shelter, if possible.

How the Fire Shelter Works

The shape of the fire shelter allows the firefighter to lie flat against the ground. This allows the firefighter to expose less of his or her body to radiant heat and more to ground cooling. The best position for a firefighter to be in is to breathe cool, clean air by pressing the face to the ground (and scooping out as much dirt as possible). The fire shelter is equipped with hold-down straps and a turned-in skirt around the edges to aid the firefighter in holding a fire shelter tightly to the ground.

If there is little or no wind, the firefighter stands inside the opened shelter with his or her back toward the oncoming flame front. The firefighter then falls to the knees before falling forward into the fully prone position **(Figure 2.14)**. Using the extremities to secure the edges of the shelter, the firefighter lies face down with the feet toward the fire **(Figures 2.15 a and b)**. The foot end will be the hottest spot in the shelter but is easily held down with boots.

Figure 2.14 The firefighter stands inside the opened shelter with his or her back toward the oncoming flame front and then falls to the knees.

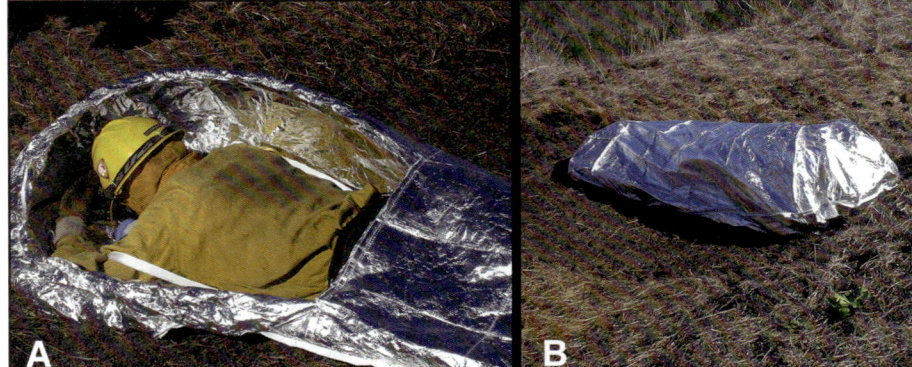

Figures 2.15 a and b The firefighter falls forward into the full prone position and uses the extremities to secure the edges of the shelter. b) The firefighter lies face down with the feet toward the fire.

Firefighters should be aware that the shelter cannot protect everyone in all circumstances, although new generation shelters offer significantly more protection from radiant heat and direct flame. As previously stated, firefighters' highest priority should always be to avoid situations that can lead to entrapment.

Determining Fire Shelter Location

Effective fire-shelter deployment depends on recognizing the need for the shelter soon enough to allow time to get to a safety zone and deploy the shelter **(Figure 2.16)**. In most cases, the crew leader or supervisor decides when and where firefighters deploy their shelters. If firefighters are not part of a crew or if they become separated from their crew, they must make these decisions for themselves. Firefighters will improve their chance of survival if their deployed shelter will not be exposed to direct flame.

Figure 2.16 In most cases, the crew leader decides when and where firefighters deploy their shelters.

Choose a site where the following are present:

- Light fuels
- Natural firebreaks, such as:
    - Creek beds
    - Depressions in the ground
    - Rock slide areas — rocks should be small enough that your shelter is still able to rest firmly on the ground
    - Lee side of ridge tops and hills
    - Flat areas on slopes, such as benches or road cuts
- Wide control lines, such as dozer lines
- Burned areas with no reburn potential
- Areas where the flame front will pass quickly

Do not place fire shelters where the following factors are present:

- Thick vegetation, such as tall grass, small trees, trees with low branches, or brush.

- Trees, logs, and snags or anything that will fall on firefighters.
- Areas, such as draws, saddles, or chimneys, where flame will race up.
- Roads where there is traffic.
- Rock slide areas where you are not able to keep the edges of the shelter firmly on the ground.

The following items should NOT be anywhere near your deployed shelter:

- Fusees
- Gasoline cans
- Supply boxes
- Packsacks
- Other combustible fire fighting gear

If firefighters must deploy their shelters in windy conditions, they should lie on their backs with their heads toward the wind. Holding the top of the shelter, they should allow the wind to fill the shelter with air as they insert their boots inside the straps at the bottom of the shelter. When the bottom of the shelter is secure, they pull the rest of the shelter down over themselves. Then they carefully roll over inside the shelter so that they can lie face down.

Agency Policy

Every fire department should have established policies and procedures relating to the maintenance and use of personal protective equipment, including fire shelters. These policies and procedures should be in compliance with applicable laws and standards, such as NFPA® 1051, *Standard for Wildland Firefighting Professional Qualifications* and with accepted safe practices within the fire protection community. These procedures should emphasize that firefighters must not be in an IDLH environment, such as a wildland fire, without proper PPE, an assignment, and a supervisor. Firefighters are responsible for knowing and complying with their agency's policies and procedures.

Rules of Engagement

Safety is about decision-making and knowing and understanding fire behavior. The decision to engage in suppression operations or when to disengage and reassess a situation is based on timely information and trigger points. Ahead of initial operations, outline these trigger points based on existing criteria and current and expected fire behavior. Following are some standardized criteria developed over time and tested through experiences encountered by thousands of firefighters.

Situational Awareness

Situational awareness (SA) is the foundation for decision-making in the ground cover fire environment. As a firefighter, you will be expected to receive and follow instructions from a supervisor, crew leader, company officer, or Incident Commander (IC). Therefore, it is imperative for you to maintain situational awareness throughout the entire operation. A large part of situational awareness is communicating what you see to your immediate supervisor, crew leader, company officer, or IC. New information about conditions may lead them to change their orders and strategies during an incident. Employ the fundamental principle of situational awareness while working on an incident and detect problems early.

Communicating observations when assigned a task is essential for safety. In addition to those observations, the following conditions should be observed and communicated:

- Fuel characteristics
    — Fuel type such as grass, trees, or underbrush
    — Fuel moisture content (wet or dry)
    — Arrangement of fuels and location of exposures
- Topography
    — Location of the fire
    — Upward slope or downward slope
    — Terrain features (canyons, chutes)
- Weather
    — Visibility
    — Wind speed and direction
    — Relative humidity
    — Thunderstorm development or other weather events
- Fire behavior
    — Location of the fire
    — Changes to spread, and growth
- Location of resources
    — Water supply
    — Personnel
    — Apparatus/equipment

It can be a challenge for firefighters to maintain their situational awareness on the fireline. The following are risks to situational awareness:

- Inexperience
- Job-related stress
- Fatigue
- Personal and environmental distractions
- Attitude

Look Up, Look Down, Look Around

A habit that all firefighters should develop is to continually check their environment: *Look Up, Look Down, Look Around*. Developing this habit is one of the best ways firefighters can protect themselves in hazardous situations.

Look Up

When firefighters look up, they may see any number of potential hazards (**Figure 2.17**). The following are critical to firefighter safety.

Figure 2.17 It is critical for firefighters to LOOK UP.

- When crews or equipment are working upslope, firefighters below must be aware that rocks may be dislodged and roll down the hill toward them or that burning materials may roll into unburned fuel below them.
- Overhead power lines may burn through and fall on anyone below.
- Staying aware of what aircraft flying overhead are doing can reduce the chances of being hit by an airdrop.
- During mop-up, firefighters must be wary of tree limbs weakened by fire and burning or burned-out snags that may fall without warning.

Look Down

Firefighters should always be careful about where they step **(Figure 2.18)**. For example, poor footing can cause firefighters to fall, especially on steep slopes. By looking down, firefighters should notice the following:

- **Downed power lines**. Stepping on a downed power line can be fatal.
- **Ditches, holes, or drop-offs**. Firefighters who walk into them can be injured or killed.
- **Snakes**. Snakebite is a possibility.
- **Fuels that cause changes in fire behavior**. Observe the fuels to anticipate changes in fire behavior.
- **Fire downslope from your position**. Observe carefully and often a fire downslope from your position.

Look Around

In the sometimes noisy environment of the fireline, firefighters may not hear signs of danger approaching. Someone operating a chainsaw may not hear a vehicle approaching. Looking around can alert firefighters to approaching vehicles as well as to changes in the fire's behavior that might threaten them. Firefighters should also look around when working on the fireline to maintain 10 feet (3 m) between themselves and other crew members.

Avoiding Lightning

Whenever lightning strikes reach the earth (ground flashes), firefighters in the area are at risk of being struck. The thunder it produces is less than 30 seconds; therefore, firefighters should take precautions against being struck. The "area" can be as much as 40 miles (60 km) ahead of a thundercloud formation. However, most ground flashes occur directly below a cumulonimbus cloud.

The U.S. Forest Service (USFS) has identified several common fire behavior characteristics that have resulted in firefighter fatalities or near misses. The study emphasized that firefighters must:

- Remain alert for potentially life-threatening situations, even when a fire does not appear to be dangerous.
- Monitor both the fire's behavior and the factors that could modify the fire's behavior.

Watch for signs of situations developing and be prepared to modify plans accordingly. For example:

Figure 2.18 Firefighters must pay attention to where they step as they walk.

- If the fire is burning in light fuels (such as grass, herbs, and light brush), these fuels are more responsive to changes in weather conditions than are heavier fuels.
- Unexpected shift in wind direction or wind speed.
- When the fire responds to topographic conditions and runs uphill.

30/30 Rule

As a rule-of-thumb, when the interval between a ground flash and the thunder it produces is less than 30 seconds, firefighters should take precautions against being struck. These precautions should be continued for 30 minutes after the thundercloud passes. Under these conditions, firefighters should avoid the following:

- Dry creek beds because of the potential for flash floods
- Being close to heavy machinery
- Being close to flammable liquids
- Poles, trees, ridge tops, ledges, wire fences, and rock outcroppings
- Lying down
- Having metal tools nearby
- Using landline telephones
- Using radios that have long antennas

In addition, to protect themselves from lightning strikes, firefighters should do the following:

- Get inside of a vehicle or structure if possible.
- Get in the middle of a large clearing.
- Sit on their packs or crouch with feet together (especially if skin tingles).
- Use only cell phones or radios that have short antennas.

Common Denominators of Fire Behavior on Tragedy Fires

The U.S. Forest Service (USFS) has identified several common fire behavior characteristics that have resulted in firefighter fatalities or near misses. These characteristics are called the *Common Denominators of Fire Behavior on Tragedy Fires*. The study emphasized that firefighters must remain alert for potentially life-threatening situations, even when a fire does not appear to be dangerous. Firefighters on the line must monitor both the fire's behavior and the factors that could modify the fire's behavior. Firefighters must consider the following:

- Most incidents happen on the smaller fires or on isolated portions of larger fires.
- Most fires are innocent in appearance before the "flareups" or "blow-ups." In some cases, tragedies occur in the mop-up stage.
- Flare-ups generally occur in deceptively light fuels.
- Fires run uphill surprisingly fast in chimneys, gullies, and on steep slopes.

- Some suppression tools, such as helicopters or air tankers, can adversely affect fire behavior. The blasts of air from low flying helicopters and air tankers have been known to cause flare-ups.

> **NOTE:** To reevaluate tactics and safety considerations during the afternoon, consider alignment of topography and wind a trigger point.

Wildland Fire Apparatus Safety

Wildland fire apparatus safety consists of considerations for the safety of the personnel operating on and around the equipment and for the protection of the apparatus from mechanical damage and exposure to the fire. Apparatus safety can be broken down into the following broad categories:

- General guidelines
- Off-road guidelines
- Engine operations
- Safety and personnel transport

General Guidelines

When responding to, working on, or returning from a fire, apparatus operators are responsible for the safe operation of the vehicle and for the safety of the personnel on and around the vehicle, including pedestrians. Part of your role on any scene is to operate equipment in a way that protects you and the others around you.

The following are some commonsense safety rules for drivers and passengers:

- **Drive at safe speeds**. Saving structures or vegetative fuels does not justify having an accident.
- **Ride inside the vehicle**. Do *not* ride on the tailboard, running boards, bumpers, fenders, or any other area.
- **Wear safety belts and protective clothing at all times**.
- **Turn on fire apparatus headlights — day or night**. Whenever the vehicle is in motion, the headlights must be on at all times.
- **Stay awake and alert**. If you are sleepy, do not drive!

Poor visibility can be an issue in smoke, fog, or darkness. Vehicle operators must slow down, and use a spotter to watch for obstacles that may be encountered at a ground cover fire. Spotters should be as visible as possible to the driver by carrying lights, wearing appropriate clothing, and staying in the driver's field of view. Also, it is important for the spotter and driver to agree on a set of hand signals for common maneuvers such as turning, stopping, and reversing. The spotter should get in the habit of using his or her arms to show the decreasing distance to objects like trees and rocks. First and foremost, the operator should reduce speed appropriately.

Off-Road Guidelines

Fighting ground cover fires often requires driving your apparatus off the roadway to reach the fire. Some fire apparatus are specifically designed for off-road use, and some are not. For example, some fire apparatus are not suitable for off-road use because of their weight, low ground clearance, and a large turning radius.

Capabilities and Limitations

You have to know what you can do safely — even if you are driving a heavy-duty brush truck. If you mishandle or overtax your vehicle, you can get yourself killed or damage expensive equipment. Even with apparatus specifically designed for off-road use, apparatus operators must know their units' capabilities and limitations and operate them accordingly. Damage to the vehicle and injury to personnel can result from mishandling or overtaxing the vehicle.

The speed at which a vehicle is driven should reflect the driving conditions. The crew and apparatus could be in jeopardy if a breakdown occurs. Also, you may be operating in hilly terrain—don't forget to use the transmission to slow down on steep descents. Brake overheating and failure are common problems in these situations.

Hoseline Safety Guidelines

Consider these guidelines for hoseline safety:

- Use hose bed covers to protect equipment stored in open hose beds.
- Use hose bed covers to protect hose from embers.
- Connect a protection line for rapid deployment, and have it charged and ready.
- Staff engines with at least three (3) personnel, including a driver/operator, a nozzle operator, and at least one additional firefighter.
- Wear eye protection if you are a nozzle operator.
- Keep hoselines as short as possible.
- Do not block access ways with hoselines.
- Lay supply lines on the shoulder of the road only.

In addition, driver/operators should remember that even though all-wheel-drive or four-wheel-drive vehicles have superior climbing ability, they also have a higher center of gravity, which makes them more susceptible to rollover than other types of apparatus. Consider the safe slope limitations mentioned earlier when operating these vehicles.

Cautions in Off-Road Apparatus Operation

When operating off-road, maneuvering an apparatus can be difficult. Even though all-wheel-drive or four-wheel-drive vehicles have superior climbing ability, they also have a higher center of gravity that makes them more susceptible to rollover than other types of apparatus. Be aware of the following areas when operating off-road:

- **Loose and unstable ground**. On steep hillsides, loose or unstable ground can cause the vehicle to slide or overturn, especially if the vehicle has a relatively high center of gravity.)
- **Slopes**. Never drive up and down slopes exceeding 40 percent or across slopes exceeding 20 percent. Many newer vehicles are equipped with inclinometers to help identify these limitations.

- **Soft terrain**. Even on level terrain, a vehicle can become mired in soft ground, sand, or mud, leaving it vulnerable to being overrun by a fire.
- **Bridges and streams**. Never drive across a bridge unless you know it to be strong enough to support the vehicle's weight. And although you will be tempted, do not attempt to ford streams with apparatus that is not designed to do so.
- **Railroad bed shoulders**. The shoulders of railroad roadbeds are not designed for vehicle traffic **(Figure 2.19)**. It may look like the quickest way from point A to point B, but you can damage your tires on the coarse, angular rock used on these roadbeds. Also, this rock is loose, and you will be in danger of sliding and rolling over on these steep inclines.

Figure 2.19
Apparatus should not be driven on the slopes of railroad roadbeds.

Even all-wheel-drive or four-wheel-drive vehicles can get stuck. If you get stuck, you can't fight the fire, and you may be putting yourself, your crew, and your vehicle in the path of the fire. If you encounter any of these five situations, raise the yellow flag and take heed.

Personnel Transport

Firefighters are often transported from an incident base to a forward staging area or from one section of fireline to another in some sort of vehicle. The type of vehicle can be an engine, a helicopter, or a personnel transport unit. The basic safety rules that apply to passengers in any emergency vehicle also apply to those being transported to or from their assignments in a ground cover fire. Transporting personnel with land-based vehicles can be a relatively safe operation if the following guidelines are used:

- Drivers must be qualified for the vehicle and operating conditions. If not, they must not be permitted to drive.
- Have shifts that do not exceed the maximum duty day for your agency, with no more than 10 hours of behind-the-wheel driving.
- Get at least eight hours of off-duty time between shifts.
- Be responsible for the safe operation of the vehicle, including using wheel chocks if provided and making daily mechanical checks before driving.

- Remove unsafe equipment from service and report equipment status to the ground support unit.
- Perform a vehicle walk-around before departure.
- Use spotters outside of vehicles when backing or turning around.
- Observe all traffic signals and follow safe speed limits.
- Follow headlight and safety rules at all times.

Heavy Equipment Operations

All mechanized equipment and mobile apparatus used on ground cover fires (including contract equipment) must be equipped with seat belts, proper rollover protection, lighting, back-up alarms, fire curtains, fire shelters, and personnel protective gear before being placed in service. Operators must be trained in the use of this safety equipment and required to use it. They must also be trained in ground cover fire operations using their type of equipment **(Figure 2.20)**. Operators should be required to check all safety gear and safety features at least daily, usually during a regular morning inspection and when changing shifts or operators.

Figure 2.20 Operators must be trained in ground cover fire operations using their type of equipment.

All vehicles must have:
- A service brake system, an emergency brake system, and a parking brake system
- Working headlights, tail lights, and brake lights
- An audible warning device (horn)
- Intact windshield with working windshield wipers

Firefighters will need to know the following:
- Hazards of working around heavy equipment

- Communication with operators
- Safety procedures for firefighters

In addition to standard fire fighting apparatus, a number of different types of heavy equipment may be used to support a major ground cover fire fighting operation. This equipment is used primarily to construct firebreaks that may slow or halt the spread of an advancing fire or to provide a means of anchoring a firing operation. It is important that firefighters be familiar with the various types of equipment that are available to them and know their capabilities.

The following are the three most common types of mechanized equipment used in ground cover fire fighting:

- Dozers
- Tractor-plows
- Road graders (maintainers)

Depending on local conditions, other types of equipment may be used as well. Farm plows, large mowers, and similar types of equipment may be used to cut firebreaks if the fire is in terrain such as crop lands, pastures, and parks that is suitable for these implements. In many cases, their use is similar to the principles described for the other heavy equipment covered in this section.

Heavy Equipment Hazards

Your high priority safety concerns are seeing potential hazards and being seen by equipment operators. Be alert at all times, especially when equipment is moving. Remember to look up, look down, and look around:

- Know where the equipment is located.
- Watch for *debris* dislodged upslope. When a large rock or similar item becomes dislodged, it can roll down with sufficient speed to injure or kill anyone in its path.
- Watch for *burning items* dislodged upslope. Burning pinecones, logs, and other objects can roll down the hill and start fires below you and your crew.

Seeing and Being Seen

Because of the noise produced by running heavy equipment and by the fire, most equipment operators rely primarily on their vision to locate hazards. Never assume the operator knows where you are. To stay visible to equipment operators at all times:

- Post a lookout if your crew operates near heavy equipment.
- Have proper personal protective equipment (PPE) to increase visibility — including headlamps or chemical light sticks and reflective clothing.
- Approach carefully — night or day, approach mechanized equipment from the sides, only after being signaled to do so by the operator.

Heavy Equipment Operations Safety

Using heavy equipment also presents additional challenges when it comes to getting out of the way of a fire. Heavy equipment is usually slow moving and therefore vulnerable to being overrun by a rapidly advancing fire front. This is an obvious hazard for the operator, but it can also be a problem for you if the equipment blocks your escape route. To avoid this predicament, operators and their support personnel must pay close attention to a fire's behavior and be prepared to withdraw into safety zones if the fire threatens their positions. Periodically, operators should use the equipment to develop safety zones.

When firefighters are assigned to work in close proximity to mechanized equipment on the fireline, they are exposed to an additional set of risks. The possibility of being struck or run over by a piece of heavy equipment is always present. Firefighters and heavy equipment operators should use the following safety procedures when working around heavy equipment:

- Stay at least 100 feet (30 m) in front and 50 feet (15 m) behind the equipment. In timber, increase the distances to 2½ times the canopy height.
- Do not allow anyone but the operator to ride on the equipment.
- Maintain eye contact with the operator when approaching equipment. Ensure that all implements have been lowered to the ground and that the equipment is idled down.
- Avoid working below equipment where rolling material could jeopardize your safety.
- Use headlamp and/or glow sticks so that the operator can see you. Night work is more dangerous due to reduced visibility.
- Establish visual and radio communication methods prior to engaging.
- Communicate all hazards (spot fires, firing operations, and obstacles) to the operator.
- Take responsibility for your safety and the safety of all those around you. Equipment operators have difficulty seeing ground personnel.

Hand Tool Safety

Hand tools such as axes, Pulaskis, McLeods, and brush hooks should have smooth, well-maintained handles and sharp cutting edges, allowing firefighters to use short, sharp cutting strokes **(Figure 2.21)**. Sharp tools will reduce the need for firefighters to raise these tools above their heads. However, when the tools are transported, blade guards should cover the sharp edges of the tool blades. Hand tools should be held at the balance point and carried at the side, close to the body, and parallel to the ground, not on the shoulder. If you take a fall when you're on foot on uneven terrain, your hand tools can injure you or a coworker. Therefore, firefighters should maintain a distance of at least 10 feet (3 m) between themselves and other firefighters.

Figure 2.21 Ensure that hand tools have smooth, well maintained handles and sharp cutting edges.

Radio Communication

Communication with heavy equipment operators is a big part of successful ground cover fire fighting. Operators must both give and receive information for the safe operation of the equipment. To integrate their equipment activities into the overall plan, operators must talk to both supervisors and ground crews.

This communication happens in two ways:

- **Radios**. Because of the noise of the equipment, radio communication often requires that operators wear radios built into their hearing protection. Voice communication may be difficult or impossible except when the unit is stopped and throttled down.
- **Through the equipment itself**. Communication using engine signals.
    — Revving the motor: One way for the operator to communicate with the spotter on the ground:
    — Gunning the motor *once*: The operator wants the dozer helper to come to the dozer.
    — Gunning the motor *twice*: The operator cannot see the spotter.

Communication also happens between operators and their spotters with the equipment's motor. This communication happens in two ways:

- One rev of the engine — The dozer helper should come to the dozer.
- Two revs of the engine — The dozer operator cannot see the spotter.

Hand Signals

Use either radio communication or hand signals to save time and reduce the chance of misunderstanding when operators communicate with the ground crew. Agree upon in advance a few simple hand signals to convey these messages to aid the spotter or line supervisor in directing the activity of earth-moving equipment:

- **Stop**. Use a back and forth, waist high, swinging motion.
- **Come ahead**. Use an up-and-down motion from waist to arm's length above.
- **Turn**. Swing a flag or light on the side where the operator is to turn.
- **Reverse**. Use a full-circle arm motion in front of the spotter.
- **Caution**. Wave a flag or light in a half circle at arm's length above your head.
- **Attract operator's attention**. You may also use one blast on a police whistle or suitable substitute. Any time you need to refresh your memory on these signals, refer to the *Incident Response Pocket Guide (IRPG)*.

Safety Tips for Firefighters

As a firefighter, you have a role in maintaining everyone's safety on the fireline. When working around heavy equipment, firefighters should:

- Assess the safety of the situation continually.
- Never allow personnel to work on a slope below an operating dozer.
- Shout "Rock!" if heavy equipment working upslope dislodges rocks or other materials that start rolling downhill.
- Never try to get on or off heavy equipment while it is in motion.

Heavy Equipment and Underground Utilities

The use of heavy equipment as stated previously in this unit can be the difference sometimes between timely fire control and an extended attack operation. In remote areas the likelihood that damage will occur to underground utilities is lessoned although not negated. The probability of damage to infrastructure rises dramatically with the use of heavy equipment in the urban interface. Although most underground utilities are installed according to standard construction guidelines where they would be unaffected by removal of surface fuels by heavy equipment, bad practices and poor oversite during construction activities can be reason for conduits, pipelines and cables to be impacted by operations. Especially in the case of the use of tractor plows.

The line supervisor, swamper or heavy equipment boss needs to do a thorough job of scouting the area for utility markers and clues as to the location of underground infrastructure. If at all possible, obtain record drawings of utilities and keep them available along with other preplan information.

Below is a listing of some visual indicators that utilities are in the area:
- Valve-box covers
- Telephone and cable riser boxes
- Manhole covers
- Pipeline markers
- Pin flags
- Paint markings on hard surfaces
- Light and utility poles
- Transformer boxes
- Traffic light control panel boxes

In most states, utilities are installed along right of ways between the back of curbs or roadway edge and the edge of the sidewalk in commercial areas and the property line in residential areas if no walks are present. In some regions, utilities are installed along common property lines at the rear of the properties. Heavy equipment bosses, line scouts, equipment operators and firefighters must be constantly aware of their operational environment – **Look Up, Look Down, Look Around**. This may be the difference between safe effective operations or disaster for all.

Hazardous Materials Situations

Firefighters may have to deal with hazardous materials during a ground cover fire, especially in the wildland/urban interface. The materials may range from barrels of agricultural pesticides in farm structures to propane in large tanks at fuel distribution points. In some cases, hazardous wastes are illegally dumped in the wildland. The hazardous material may have even started the fire to which the firefighters were called.

Clandestine drug labs, often located in rural areas to reduce the chances of detection, may contain several different toxic and/or explosive chemicals. These chemicals can start or contribute to the growth of fires in the wildland. These clandestine operations are often protected by armed guards, attack dogs, and/or potentially lethal booby traps. Regardless of what the hazardous material is or how it came to be where it is, firefighters must be able to recognize and isolate it.

Hazmat Incident Operations
- Think safety.
- Assess situation.
- Safe approach; upwind/upgrade/upstream.
- Identify, isolate, establish perimeter; and deny entry.
- Notify agency dispatcher.

Scene Management
- Goal is to protect life, environment and property.
- Attempt to identify substance using the 2012 *Emergency Response Guide* (use binoculars, placards/labels, container shapes/colors, Material Safety Data Sheets, shipping papers, or license plate).
- Assess quantity of material involved.
- Identify exposures and hazards surrounding the site.
- Anticipate weather influences. Organizational responsibilities.
- Establish command including an IC and Safety Officer.
- Develop action plan for area security and evacuation.
- Keep your supervisor informed of hazards.
- Advise all on scene and responding resources of changes in situation.
- Keep dispatcher advised of changes.
- Document all actions taken.
- Make special note of any responder exposures.

Use the following guidelines:
- Identify and mark all hazardous materials with flagging material or other warning devices.
- Post a lookout.
- Avoid breathing toxic fumes unless you have been trained and equipped to use hazmat protective gear.
- Be especially cautious around petroleum and propane tanks:
    — Check all storage tanks for LPG.
    — Check for a 30 foot (10 m) vegetation clearance around storage tanks.
    — Clear vegetation if necessary and if time allows.
- Check outbuildings or barns for flammable liquid storage.
- Look for drug labs, such as those producing methamphetamines, that are often found in out of-the-way places like the wildland/urban interface. Meth labs contain several different toxic and explosive chemicals.
- Look for chemicals, pesticides, herbicides, petroleum products, and paint stored in or around the structure or in outbuildings. Remove these materials from the structure, or protect them from fire exposure.

Operations Near Electrical Power Lines

Whenever power lines are in or near a ground cover fire, all incoming units should be notified of this fact, and all incoming aircraft should be warned of the poles or transmission towers. Even power lines that have not fallen can increase the possibility of firefighters being electrocuted. High brush and small trees under power lines can sometimes cause a phase-to-ground short. Constructing a control line around the fire a distance equal to one span from the power lines is the safest tactic. Once the fire has spread away from the point of contact, fog streams may be used to extinguish the fire.

Activity near high voltage electrical transmission/distribution lines can cause multiple hazards and electrocute or seriously injure firefighters. The IC and line supervisors must be aware and communicate power line hazards to all resources. When fire crews must work in the vicinity of overhead power lines, the power company should deactivate these lines if possible.

Guidelines for Down Power Lines

- Notify all responders of down electrical lines (**Figure 2.22**). Obtain radio check-back.
- Determine entire extent of hazard by visually tracking all lines — two poles in each direction, from the downed wire.
- Flag area around down wire hazards.
- Delay fire fighting actions until hazard identification and flagging are complete.
- Downed line on vehicle: stay in vehicle until the power company arrives. If vehicle is on fire, jump out with both feet together. Do not touch the vehicle. Keep feet together and shuffle or hop away.
- Treat downed wires as energized!
- Normal tactics apply when fire is more than 100 feet (30 m) from power lines. Heavy smoke and flames can cause arcs to ground. Direct attack must be abandoned within 100 feet (30 m) of transmission lines.
- Spot fires or low ground fires can be fought with hoselines if heavy smoke or flame is not within 100 feet (30 m) of power lines.
- Maintain 35 (12 m) feet distance from transmission towers.
- Never use straight streams or foam—use a fog pattern.
- Use extreme caution if engaging in tactical firing operations.
- Extinguish wooden poles burning at the base to prevent down wire hazards.

Figure 2.22 All responders should be notified of down electrical lines.

Snag Safety

Snags are standing dead trees. They present a variety of hazards to firefighters working near them. They often smolder long after the main fire has been extinguished, so they must be cut down (felled) during mop-up. Dead limbs can break off and fall on firefighters attempting to fell a snag. A spotter is usually needed to watch for falling limbs during felling operations **(Figure 2.23)**. All personnel working nearby must be warned of the impending fall of a snag. If there is fire within the trunk, as there often is, the trunk must be opened in order to extinguish it.

Figure 2.23 A spotter watches as a snag is being felled.

Key Terms

Escape Route — A preplanned and understood route firefighters take to move to a safety zone or other low-risk area. When escape routes deviate from a defined physical path, they should be clearly marked (flagged) [National Wildfire Coordinating Group (NWCG) *Glossary of Wildland Fire Terminology*].

Fireline — Part of a control line that is scraped or dug to mineral soil; also, a general term for the area where fire fighting activities are taking place, the wildland equivalent of the term "fireground" as used in structural fire fighting.

Fire Shelter — An aluminized cloth tent that offers protection in a fire entrapment situation by reflecting radiant heat and providing a volume of breathable air (National Wildfire Coordinating Group (NWCG) *Glossary of Wildland Fire Terminology*.)

Lookout — (1) Person designated to detect and report fires from a vantage point. (2) Location from which fires can be detected and reported. (3) Fire crew member assigned to observe the fire and warn the crew when there is danger of becoming trapped.

Safety Zone — An area cleared of flammable materials used for escape in the event the line is outflanked or in case a spot fire causes fuels outside the control line to render the line unsafe. In firing operations, crews progress so as to maintain a safety zone close at hand allowing the fuels inside the control line to be consumed before going ahead. Safety zones may also be constructed as integral parts of fuel breaks; they are greatly enlarged areas which can be used with relative safety by firefighters and their equipment in the event of blowup in the vicinity (National Wildfire Coordinating Group (NWCG) *Glossary of Wildland Fire Terminology*).

Turn Down — A situation where an individual has determined that he or she cannot undertake an assignment as given and is unable to negotiate an alternative solution.

References

Fire Shelters - The National Interagency Fire Center (NIFC). Accessed online. https://www.nifc.gov/fireShelt/fshelt_main.html

National Interagency Fire Center (NIFC). 2017. Accessed online. https://www.nifc.gov/

National Wildfire Coordinating Group (NWCG) *Glossary of Wildland Fire Terminology*.

National Wildfire Coordinating Group (NWCG) *Incident Response Pocket Guide* (IRPG), January 2014, PMS 461. Accessed online. https://www.nwcg.gov/publications/461

National Wildfire Coordinating Group (NWCG) *Wildland Fire Incident Management Field Guide, PMS 210, April 2013*. Accessed online. https://www.nwcg.gov/publications/210

Chapter **3**

National Incident Management System – Incident Command System (NIMS-ICS)

Table of Contents

NIMS-ICS Organizational Functions......... 60

    Command Section 61

    Incident Commander (IC).......................... 62

        Deputy Incident Commander 63

        Incident Safety Officer (ISO)............................... 63

        Public Information Officer (PIO) 64

        Liaison Officer 64

        Incident Command Post (ICP) 64

    Operations Section — Operations Section Chief... 64

    Planning Section — Planning Section Chief.... 65

    Logistics Section — Logistics Section Chief..... 66

    Finance/Administration Section — Finance/Administration Section Chief................................ 67

Unified Command... 68

Incident Management 68

    Delegation of Authority............................. 68

    Organizational Principles 69

        Scalar Structure 69

        Unity of Command 70

        Span of Control.. 70

        Division of Labor 70

    Management by Objectives..................... 71

        Size-Up 71

        Incident Strategic Goals and Tactical Objectives... 72

    Resource Management 72

        Resource Terminology.. 72

        Aid Agreements 73

        Anticipating Resource Needs............................. 74

    Incident Communications 74

        Formal Communications 75

        Informal Communications 75

        Briefings... 75

    Incident Action Plan (IAP) 76

        Verbal IAP 77

        Assigned Incident Tasks 77

        NIMS IAP Planning Process 77

Key Terms 78

References.. 78

National Incident Management System – Incident Command System (NIMS-ICS)

Chapter 3

An Incident Management System (IMS) is a management framework used to organize emergency incidents. As an emergency responder, you will initiate and operate under the IMS that your AHJ uses anytime that you respond to an emergency. The IMS provides the command structure, and management terminology will be used at all emergency incidents.

By mandate, all emergency service organizations in the United States use the National Incident Management System Incident Command System (NIMS-ICS). NIMS-ICS is designed to be applicable both to small, single-unit incidents that may last a few minutes and complex, large-scale incidents involving several agencies and mutual-aid units that possibly may last days or weeks. NIMS-ICS combines command strategy with organizational procedures. It provides a functional, systematic organizational structure that clearly shows the lines of communication and **chain of command**. NIMS-ICS provides the following to an incident:

- Modular organization
- Manageable span of control
- Organizational facilities such as a command post and staging areas
- Standardized position titles
- Integrated communication
- Accountability of resources

Firefighters can use ICS for any type of incident or event (ground cover fire, structure fire, hurricane, snowstorm, flood, civil disturbance, parade, festival, or training activity). The **Incident Command System** is modular in its organization and provides the following:

- The ability to expand and contract depending on the type, size, and complexity of the incident.
- A common operating picture. It has an integrated, standardized planning process and documentation format.
- Common terminology relative to personnel, resources, and facilities.
- A manageable span of control of resources.

- The use of a Unified Command Structure. ICS allows agencies with different legal, geographic, and functional authorities and responsibilities to work together effectively without affecting individual agency authority, responsibility, or accountability.

> All operational and support personnel from every jurisdiction are required to take ICS-100 and 200 training. For training materials or to take the training online refer to: www.training.fema.gov/EMIWEB/IS/ansreq.asp
>
> All personnel at the officer level or above (or anyone serving in a single resource or Command capacity) are required to complete ICS-300 training. Chiefs and Agency Administrators or those representatives working in a Command or Multiagency Coordination system capacity are required to complete ICS-400 and Command and General staff training. Information for each level of training and frequently asked questions can be found at https://www.fema.gov/nims-frequently-asked-questions

NIMS-ICS Organizational Functions

NIMS-ICS uses specific terminology to describe levels of command and activity at an incident as follows:

- **Command**
    — Directs, orders, and/or controls resources by virtue of explicit legal, agency, or delegated authority.
    — Denotes the organizational level that is in Command (Incident Commander [IC]) of the incident.
    — Makes the lines of authority clear to all involved.
    — Follows lawful commands immediately.
- **Command Staff**
    — Public Information Officer, Safety Officer, and Liaison Officer report directly to the Incident Commander.
- **General Staff**
    — Incident management personnel who represent the major functional Sections.
- **Section**
    — Operations
    — Planning
    — Logistics
    — Finance/Administration
    — Information and Intelligence
- **Branch**
    — Responsible for functional/geographic major segments of incident operations.
    — Organizationally located between Section and Division or Group.

— Identified by Roman numeral or functional area (such as Command, Operations).

- **Division**
  - Responsible for operations within a defined geographic area.
  - Organizationally between Branch and single resources, task force, or strike team.
  - Resources (assigned to a Division) report to that division supervisor.

- **Group**
  - Organizational level, equal to Division, responsible for a specified functional assignment at an incident (such as ventilation, salvage, water supply) without regard for a specific geographical area.
  - Available for reassignment when the assigned function has been completed.

- **Unit**
  - Organizational level within a Section.
  - Fulfills specific support functions, such as the resources, documentation, demobilization, and situation units within the Planning Section.

- **Task Force**
  - Different types of resources with common communications and a pre-established leader.
  - Sent to an incident or formed at an incident to accomplish a mission or to meet a specific incident objective.

- **Strike Team**
  - The same kind and type of resource with common communications and a leader that may be pre-established.
  - Sent to an incident or formed at an incident to accomplish a mission or to meet a specific incident objective.

- **Single Resource**
  - An individual piece of equipment and its personnel complement.
  - An established crew or team of individuals with an identified work supervisor.
  - A single individual who can be used on an incident.

Command Section

Command has the delegated authority to direct, order, and control resources **(Figure 3.1)**. Lines of authority must be clear to all, and lawful commands should

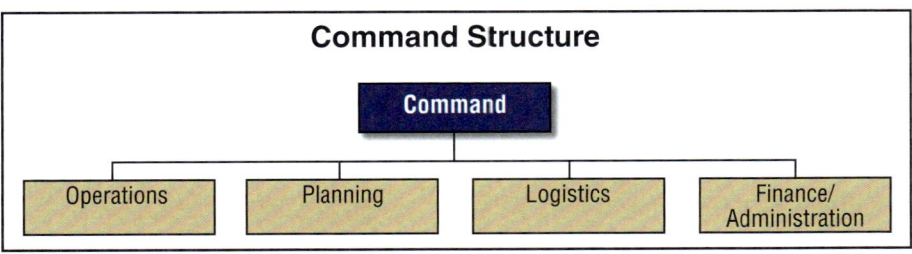

Figure 3.1 The basic NIMS-ICS Command structure.

be followed immediately without question. Responders should follow the chain of command and use correct radio protocols. To help avoid confusion during rapidly evolving situations, responders should not address anyone by name, rank, or job title; therefore, it does not matter who answers their radio messages.

The basic Command organization configuration includes the following three levels:

- **Strategic** — Entails the overall direction and goals of the incident
- **Tactical** — Identifies the objectives that the tactical level supervisor/officer must achieve to meet the strategic goals
- **Task** — Describes the specific tasks needed to meet tactical-level requirements and assigns these tasks to operational units, companies, or individuals

Incident Commander (IC)

The **Incident Commander (IC)** is the officer at the top of an incident chain of command and in charge of the incident. The IC sets the strategic goals and tactical objectives. The IC may delegate responsibilities as an incident requires. Any responsibilities not delegated remain with the IC. The Incident Commander's roles can include:

- Keeping an up-to-date report for the emergency scene
- Establishing the Command Post (CP) and formulating the **Incident Action Plan (IAP)**
- Coordinating and directing all incident resources to implement the plan and meet its goals and objectives
- Informing the telecommunicator and other responders when Command is assumed or transferred
- Establishing the site safety plan
- Implementing a site security and control plan to limit the number of personnel operating in the control zones
- Designating a Safety Officer
- Implementing appropriate emergency operations
- Ensuring that all emergency responders wear appropriate personal protective equipment (PPE) in restricted zones
- Implementing post-incident emergency response procedures (incident termination)

An aggressive plan should not be undertaken unless sufficient information is available to make logical decisions and the safe coordination of operations can be accomplished. If the incident is large and/or complex, the IC may delegate authority to the following **command staff** positions:

- Deputy Incident Commander
- Incident Safety Officer (ISO)
- Liaison Officer
- Public Information Officer

Deputy Incident Commander

At complex incidents where the IC's **span of control** is too broad, the IC may assign another officer as a deputy. The officer assigned as deputy must be qualified to assume command of the incident if needed on a permanent or temporary basis. The Deputy Incident Commander may be designated to:

- Perform specific tasks that the IC delegates.
- Relieve the IC as needed at lengthy incidents.
- Assume command if needed.
- Represent other assisting agencies that share jurisdiction.

Incident Safety Officer (ISO)

The Incident Safety Officer (ISO) is responsible to:

- Identify and monitor hazardous and unsafe situations.
- Ensure operational and personnel safety.

Although the ISO may exercise emergency authority to stop or prevent unsafe acts when immediate action is required, the officer generally chooses to correct them through regular lines of authority. The ISO must be trained to the level of operations conducted at the incident and is required to perform the following duties:

- Obtain a briefing from the IC.
- Review IAPs for safety issues.
- Identify hazardous situations at the incident scene.
- Participate in the preparation and monitoring of incident safety considerations, including medical monitoring of entry team personnel before and after entry **(Figure 3.2)**.
- Maintain communications with the IC, and advise the IC of deviations from the incident safety considerations and of any dangerous situations.
- Alter, suspend, or terminate any activity that is judged to be unsafe.
- Conduct safety briefings.
- Ensure that safety briefings are conducted for entry team personnel before entry **(Figure 3.3)**.

Figure 3.2 The Incident Safety Officer (ISO) monitors the scene for unsafe conditions.

Figure 3.3 The ISO conducts safety briefings with entry team members before entry.

Safety briefings include the following information about the incident:

— Identification of hazards

— Description of the site

— Tasks to be performed

— Anticipated duration of the tasks

— PPE requirements

— Monitoring requirements

— Notification of identified risks

— Additional pertinent information

Public Information Officer (PIO)

The Public Information Officer (PIO) is responsible for relaying accurate information between the IC and all stakeholders during and after an incident. Stakeholders could include other response sections, the media, or the public. Additional PIO duties include:

- Advising the IC on information dissemination and media relations
- Obtaining and providing information to the Planning Section, community, and media
- Relaying information that the IC has approved

Liaison Officer

The Liaison Officer communicates between Command Staff (IC and others) and supporting agencies assisting at the incident. The Liaison Officer also:

- Reports to the Incident Commander and is a member of the Command Staff
- Responds to requests from incident personnel who need to contact assisting and cooperating agencies
- Monitors operations with an emphasis on potential problems between response agencies
- Provides briefings to supporting organizations and answers their questions

Incident Command Post (ICP)

The ICP should be established at a safe location. The IC must be accessible (either directly or indirectly), and an ICP ensures this accessibility. An ICP can be a predetermined location at a facility, a conveniently located building, or a radio-equipped vehicle located in a safe area. Ideally, the ICP location will allow the IC to observe the scene, although such a location is not absolutely necessary. The location of the ICP is relayed to the telecommunicator/dispatcher and emergency responders. An ICP needs to be readily identifiable with common identifiers such as:

- Custom-designed command vehicles or removable vehicle signage
- Marked building or tent
- Pennants, flags, or signs
- Marking lights, such as vehicle hazard lights

Operations Section — Operations Section Chief

The Operations Section directly manages all incident tactical activities, the tactical priorities, as well as the safety and welfare of personnel working in the Operations Section **(Figure 3.4)**. One of the functions of the Operations section is the establishment and maintenance of the **Staging Area**. The Staging Area

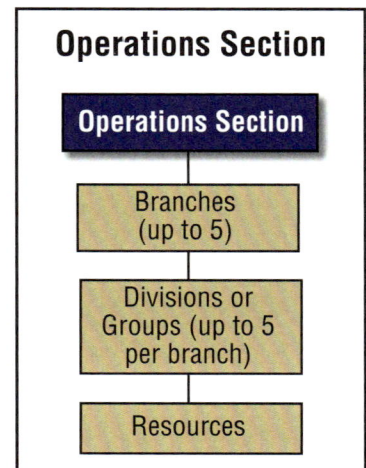

Figure 3.4 The Operations Section.

is where personnel and equipment awaiting assignment are held. This practice keeps the responders and their equipment a short distance from the scene until they are needed and minimizes confusion at the scene.

The Operations Section Chief reports directly to the IC and is responsible for managing all operations that directly affect the primary mission of eliminating a problem incident. The Operations Section Chief is responsible for:

- Directing the tactical operations to meet the IC's strategic goals
- Developing the operations portion of the Incident Action Plan
- Requesting additional resources to support tactical operations
- Approving release of resources from active operational assignments
- Making or approving expedient changes to the IAP
- Maintaining close contact with the IC, subordinate Operations personnel, and other agencies involved in the incident

Planning Section — Planning Section Chief

The Planning Section gathers, assimilates, analyzes, and processes information needed for effective decision-making **(Figure 3.5)**. Information management is a full-time task at large incidents. The Planning Section serves as the IC's clearinghouse for incidents, allowing the IC's staff to provide information instead of having to deal with dozens of information sources. Command uses the information compiled by Planning to develop strategic goals and contingency plans. Specific units under Planning include:

- Resources Unit
- Situation Unit
- Documentation Unit
- Demobilization Unit
- Technical specialists as needed

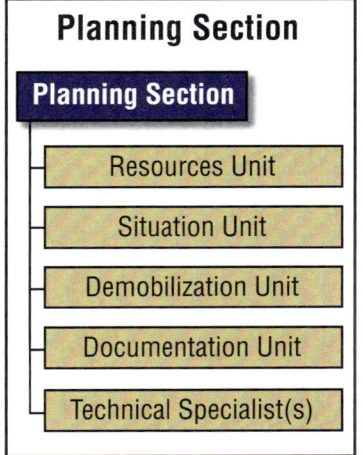

Figure 3.5 The Planning Section.

Under the direction of the Planning Section Chief, the Planning Section collects situation and resource status information, evaluates it, and processes the information for use in developing action plans. Dissemination of information can be in the form of the IAP, in formal briefings, or through map and status board displays.

The Planning Section Chief:

- Collects and manages all incident relevant operational data.
- Supervises preparation of the IAP.
- Provides input to the IC and Operations in preparing the IAP.
- Incorporates traffic, medical, and communications plans and other supporting materials into the IAP.
- Conducts and facilitates planning meetings.
- Reassigns personnel within the ICS organization.
- Compiles and displays incident status information.
- Establishes information requirements and reports schedules for units (e.g., resources, situation units).

- Determines need for specialized resources.
- Assembles and disassembles task forces and strike teams not assigned to Operations.
- Establishes specialized data collection systems as necessary (e.g., weather).
- Assembles information on alternative strategies.
- Provides periodic predictions on incident potential.
- Reporst significant changes in incident status.
- Oversees preparation of the Demobilization Plan.

Logistics Section — Logistics Section Chief

Logistics provides services and support systems to all the organizational components involved in the incident, including the following:
- Facilities
- Transportation needs
- Supplies
- Equipment
- Maintenance
- Fueling supplies
- Meals
- Communications
- Responder medical services

The Support Branch and Service Branch are two Branches within the Logistics Section **(Figure 3.6)**. The Service Branch includes medical, communications, and food services. The Support Branch includes supplies, facilities, and ground support (vehicle services).

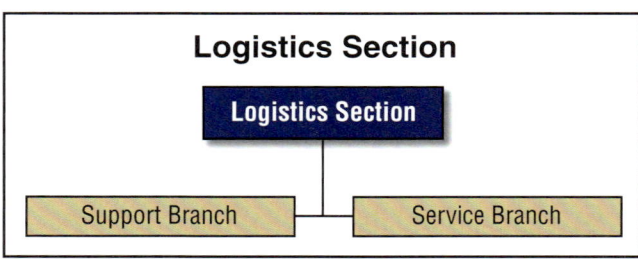

Figure 3.6 The Support Branch and the Service Branch are two branches within Logistics.

The Logistics Section Chief:
- Manages all incident logistics.
- Provides logistical input to the IAP.
- Briefs Logistics staff as needed.
- Identifies anticipated and known incident service and support requirements.
- Requests additional resources as needed.
- Ensures and oversees the development of the Communications, Medical, and Traffic Plans as required.
- Oversees demobilization of the Logistics Section and associated resources.

Finance/Administration Section — Finance/Administration Section Chief

The Finance/Administration Section is established when agencies responding to incidents have a specific need for financial services **(Figure 3.7)**. Not all agencies require the establishment of a separate Finance/Administration Section. In some cases, such as cost analysis, that function could be established as a Technical Specialist in the Planning Section. Specific units under the Finance/Administration Section include:

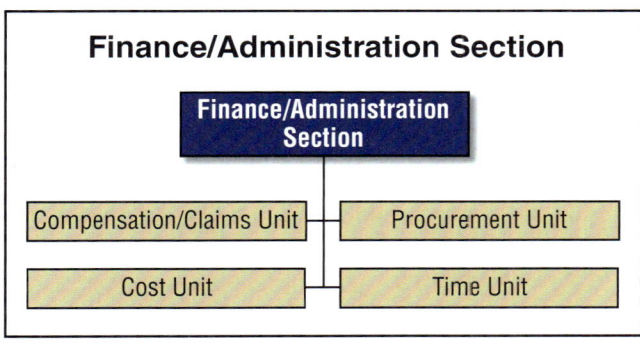

Figure 3.7 At large-scale, long-term incidents, the Finance/Administration is often activated.

- Time Unit
- Procurement Unit
- Compensation Claims Unit
- Cost Unit

The Finance/Administration Section Chief:

- Manages all financial aspects of an incident.
- Provides financial and cost-analysis information as requested.
- Ensures compensation and claims functions are being addressed relative to the incident.
- Gathers pertinent information from briefings with responsible agencies.
- Develops an operating plan for the Finance/Administration Section and fill Section supply and support needs.
- Determines the need to set up and operate an incident commissary.
- Meets with assisting and cooperating agency representatives as needed.
- Maintains daily contact with agency headquarters on finance matters.
- Ensures that personnel time records are completed accurately and transmitted to home agencies.
- Ensures that all obligation documents initiated at the incident are properly prepared and completed.
- Briefs agency administrative personnel on all incident-related financial issues needing attention or follow-up.
- Provides input to the IAP.

Unified Command

A multijurisdictional incident involves services (such as fire, law enforcement, and EMS) that are beyond the jurisdiction of one organization/agency. In these situations, the chain of command must be clearly defined. Control of an incident involving multiple agencies with overlapping authority and responsibility is accomplished through the use of **Unified Command**. When working under a Unified Command structure, several individuals may be working in Command, but only one person will ultimately answer for the operation. Unified Command simply means that all agencies that have a jurisdictional responsibility at a multijurisdictional incident contribute to the process. The following actions should be taken:

- Determine overall incident objectives.
- Select strategies.
- Accomplish joint planning for tactical activities.
- Ensure integrated tactical operations.
- Use all assigned resources effectively.

Proactive organizations identify target hazards in their areas of jurisdiction and also identify any other agencies with authority and responsibility for those target hazards. Ideally, those agencies meet, identify differences in agency IMS practices, and establish a **Memorandum of Understanding (MOU)** for Unified Command: A written agreement defining roles and responsibilities within a Unified Command structure. The lead officials of the agencies sign the MOU, and it becomes policy governing personnel from all agencies.

Proper planning and preparation lead to safe and successful responses to incidents. The occurrence of a serious incident is not the time to discover that a neighboring fire department or industry cannot provide necessary equipment, personnel, or technical expertise. When emergency services organizations work together to develop their preincident surveys, they can meet the following objectives:

- Share vital resource information.
- Develop rapport among participating emergency services organizations.
- Identify and pool needed resources.

Incident Management

Firefighters should understand how incidents are managed in order to fill various positions and also take command of an incident or lead a crew. The sections that follow provide information on incident management information for commanding incidents according to NIMS-ICS.

Delegation of Authority

In most jurisdictions, the authority for protecting citizens rest with the elected officials. The fire department's authority to respond to emergencies is an extension of that authority. Command at a scene falls within the authority to protect the public and extends from the IC to the other supervisors at the scene.

Firefighters have authority to protect the public, but that authority has limits. Firefighters must act in accordance with existing laws and agency policies and procedures. Depending upon the incident, the AHJ may also place limits on their authority specific to a particular incident.

ICs may need to have additional authority designated to them when:

- The incident is outside the IC's jurisdiction.
- The incident scope is complex or beyond existing authorities.
- There are legal requirements for a designation of authority.

A delegation of authority should include transfer of all command functions such as:

- Legal authority
- Reporting requirements
- Process for communications
- Planning for the ongoing incident

Issuance of a delegation of authority may also take into consideration political implications and agency or jurisdictional priorities.

Organizational Principles

Effective incidents are managed according to well-established organizational principles including the following:

Scalar Structure

The scalar structure is defined as having an uninterrupted series of steps or a chain of command. Decisions and information are directed from the top of the organizational structure down through intermediate levels to the base of the structure. Feedback and information, in turn, are transmitted up from the bottom through the structure to the top positions (**Figure 3.8**).

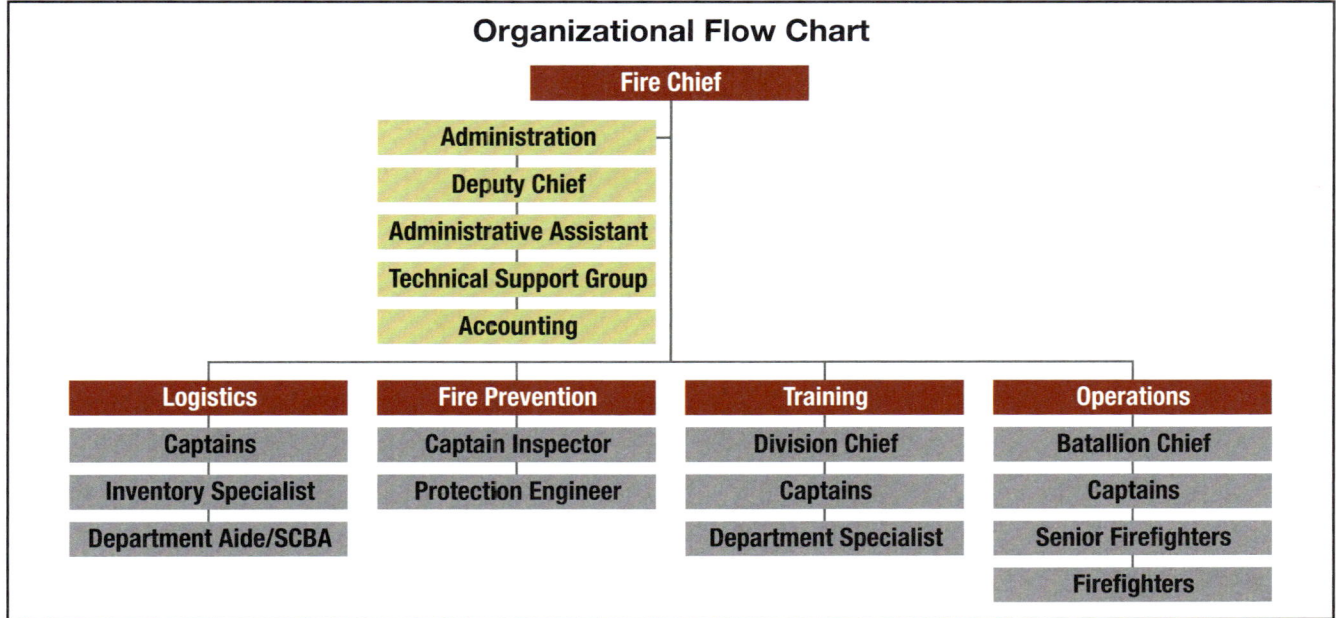

Figure 3.8 An organizational flow chart illustrating the scalar structure of a fire and emergency services organization.

Unity of Command

In Unity of Command, each firefighter or responder answers to a single supervisor during an incident. A crew member reports to the crew leader, and crew leaders report to the IC or to a Branch supervisor at larger incidents.

Span of Control

Span of control refers to the number of subordinates and/or number of functions that one individual can effectively supervise. This principle applies equally to supervising the crew of a single company or the officers of several companies under the direction of an Incident Commander. There is no absolute rule for determining how many subordinates or functions that one person can supervise effectively. The number varies with the situation but is usually considered to be somewhere between three and seven (**Figure 3.9**).

Division of Labor

Division of labor consists of dividing large jobs into smaller tasks to be assigned to specific individuals. Work groups may be created based upon the following elements:

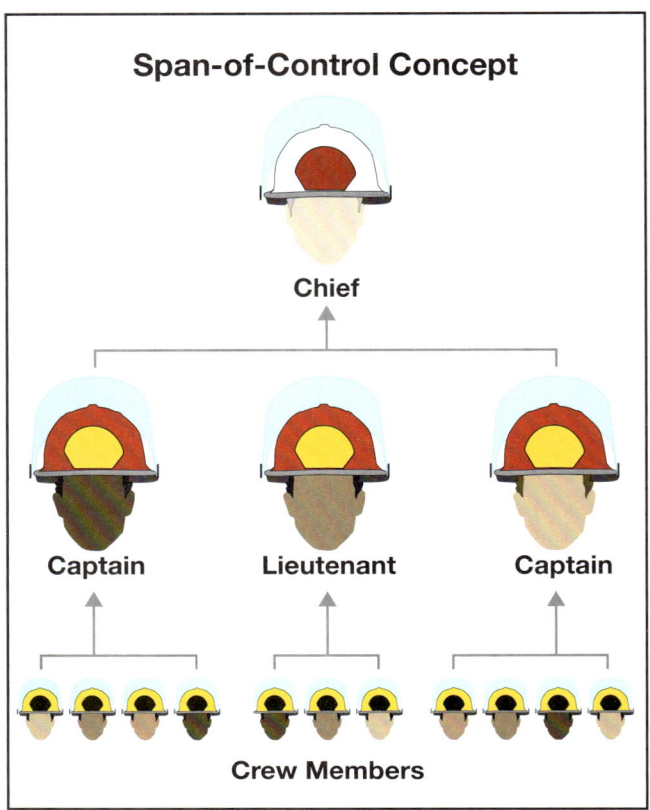

Figure 3.9 Span of control limits the number of personnel each supervisor can effectively control.

- **Type of task**. Organizations commonly place emergency work tasks into similar groups, such as engine, brush, and water supply companies; and assign personnel and equipment to handle these tasks.

- **Geographical area**. Assignments based on the arrangement of districts or battalions within the organization's response area.

- **Available resources**. Another consideration is the number of people needed to accomplish the assigned tasks.

- **Skills specialization**. Assignments should be given to the best available personnel for that particular job. An effective way of ensuring the availability of qualified personnel for anticipated assignments is to train individuals to perform particular jobs.

- **Organizational flexibility**. NIMS-ICS is built of modular Sections, Divisions, Task Forces, and/or Strike Teams. The structure is designed so that only those pieces needed at a scene are used. Similarly, if an incident becomes more complex, pieces can be easily integrated into the Command structure as needed. In addition, supervisors and section leaders can modify the structure within their areas of authority without first receiving orders or permission to do so.

> **CAUTION:**
> **Avoid combining ICS positions in an attempt to gain staffing efficiency. Units may have a common supervisor, but each unit should have its own, specific assignment.**

Management by Objectives

The Operational Planning "P" was initially developed for the U.S. Coast Guard's Oil Spill Field Operations Guide and has evolved for all-risk, all-hazard responses. This model follows a sequence of actions that are critical to using the model, although numerous simultaneous actions are involved in it **(Figure 3.10)**. The incident action plan results from this planning process.

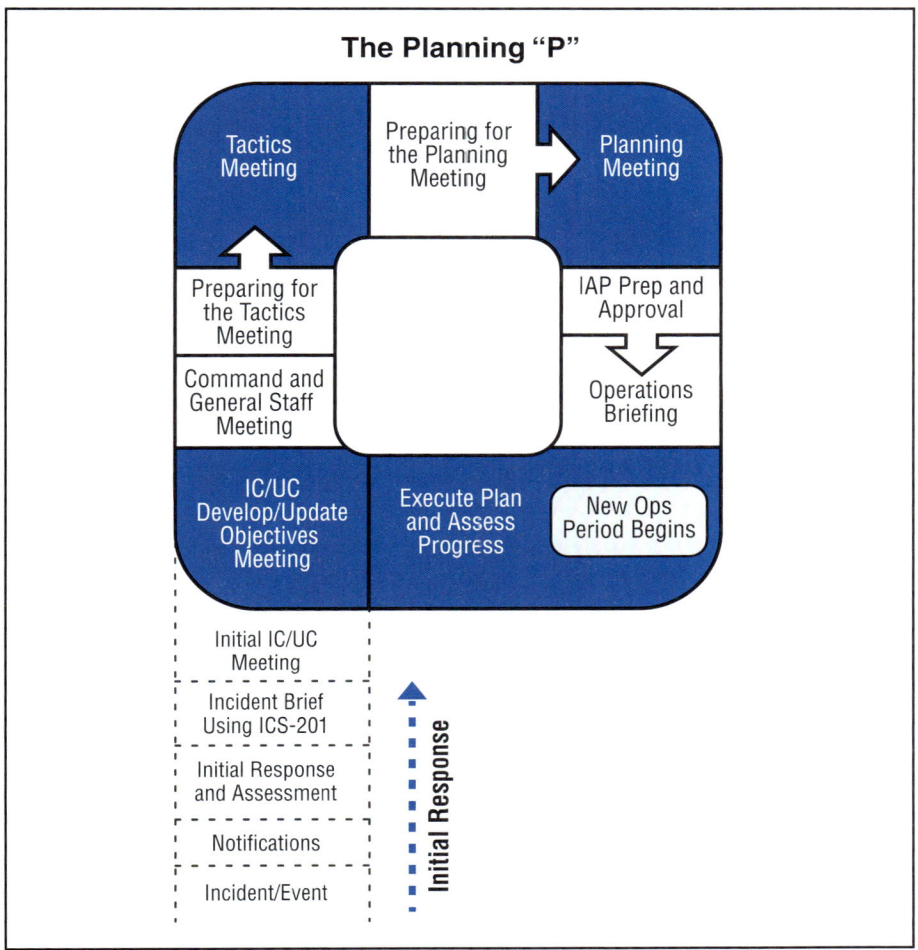

Figure 3.10 The Planning "P" model.

The Operational Planning "P" describes an ICS planning process that focuses on the first five steps of the NIMS-ICS planning process:

1. Understand the situation (size-up).
2. Establish incident objectives and strategies.
3. Develop the plan of action (IAP).
4. Prepare and disseminate the plan (make assignments).
5. Evaluate and revise the plan.

Size-Up

The size-up process actually begins before an incident is reported and continues throughout the incident. Information about an incident can be gathered from:

- Preparedness plans and preincident surveys

- Dispatch information gathered while en route
- Observations from 360-degree walkaround of the scene (if possible)
- Observations of changing conditions during the incident

Size-up is intended to identify the following:

- Size and nature of the incident.
- Hazard and safety concerns facing personnel and victims (if any).
- Areas that need to be isolated and/or secured.
- Locations of trapped, injured, or deceased occupants.
- Initial objectives and resource requirements.
- Locations for the following:
    — Control zones
    —Incident Command Post
    — Staging area
    — Entrance and exit routes for responders and civilians

Incident Strategic Goals and Tactical Objectives

The strategic goals of an incident are the overall desired outcomes. The tactical objectives are the activities used to reach those outcomes. Both are included in the IAP and must be communicated to all incident personnel.

Strategic goals are broad, general statements of the final outcomes to be achieved. The following are strategic goals that apply to all emergency situations:

- Life safety
- Incident stabilization
- Property conservation

Achieving tactical objectives leads to the completion of strategic goals. Units and personnel are assigned specific tasks to accomplish each tactical objective. Some common tactical objectives include:

- Initiate search and rescue.
- Protect exposures.

Strategic goals and tactical objectives must be regularly evaluated to ensure that they are being accomplished. As goals and objectives are met and situations change, so do the priorities.

Resource Management

Resources are needed to accomplish goals and objectives. Allocating those resources and ensuring that resources are adequate are key to successful incident management. Resources may be available from your organization or may be provided through aid agreements.

Resource Terminology

NIMS-ICS uses the following terminology to describe resources:

- **Crew.** Specified number of personnel assembled for an assignment, such as exposure protection, fire attack, and water supply operations. The number of personnel assigned to a crew should be within span-of-control guidelines. A crew operates under the direct supervision of a crew leader.

- **Single resources**. Individual pieces of apparatus (engines, brush trucks, water tenders, bulldozers, air tankers, helicopters) and the personnel required to make them functional.
- **Task force**. Any combination of resources (engines, brush trucks, bulldozers) assembled for a specific mission or operational assignment. All units in the force must have common communications capabilities and a designated leader. Once a task force's tactical objective has been met, the force is disbanded; individual resources are reassigned or released.
- **Strike team**. Set number of resources of the same kind and type (engines, brush trucks, bulldozers) that have an established minimum number of personnel. All units in the team must have common communications capabilities and a leader in a separate vehicle. Unlike task forces, strike teams remain together and function as a team throughout an incident.

Aid Agreements

Mutual, automatic, and outside aid agreements are the result of reciprocal agreements between multiple jurisdictions and organizations, both public and private (**Figure 3.11**). These agreements may be verbal but should be written. Fire and emergency services organizations commonly enter into aid agreements to:

- Receive state or federal funding.
- Share limited or specialized resources between neighboring jurisdictions.
- Assist neighboring jurisdictions when a response requirement exceeds the primary jurisdiction's capabilities.
- Meet operational requirements when its own resources are deployed at an incident and a second, simultaneous emergency occurs.
- Assist organizations in meeting NFPA®, Insurance Services Office (ISO), or Center for Public Safety Excellence department accreditation process, and other requirements for staffing, apparatus availability, response times, etc., through shared resources.
- Define responses for areas on the boundaries between adjacent jurisdictions.
- Define response methods for fire and emergency services organizations within a jurisdiction, such as a military base or corporate fire protection agency (fire brigade) within a city's limits.
- Assist neighboring jurisdictions with target hazards or high-risk facilities.
- Provide a quicker response when other fire and emergency services organizations are closer to the emergency site than are the primary jurisdiction's resources.

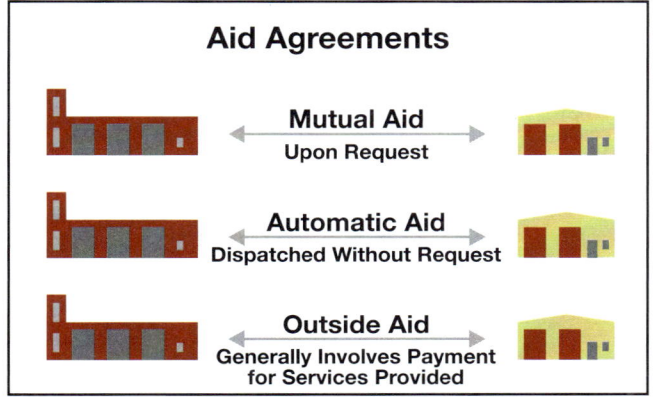

Figure 3.11 Illustrating common aid agreements between fire departments.

The implementation of new policies or ordinances may be necessary to support the aid agreements. All fire and emergency services organizations participating in these agreements should conduct joint training exercises so that differences in equipment and procedures may be identified and rectified before a major incident occurs.

Outside aid is similar to mutual aid except that payment, such as a per response or an annual fee, rather than reciprocal aid is made by one organization to the other. Outside aid agreements differ little from the other aid agreements. Outside aid agreement should define:

- Conditions under which support will be provided:
    — Automatic
    — On request
- Terms for conducting the response:
    — Command and communication
    — Standard operating procedures/standard operating guidelines (SOPs/SOGs)
    — Legal considerations, among other considerations

Anticipating Resource Needs

Experience is the best guideline for anticipating needs at an incident. There are some guidelines that are true for most incidents:

- The Operations Section has the heaviest workload and requires the most resources.
- Other sections, such as Planning, may need more resources initially but fewer resources as an incident proceeds. Planning resources could be reassigned later in the incident.
- Having too few resources can increase the risk of life and property loss.
- Having too many resources increases the risk of unqualified personnel being deployed without proper supervision.
- Scene complexity also has an effect on needed resources at the scene. Complexity factors include:
- Firefighter and occupant safety
- Weather and other environmental concerns
- Likelihood that an incident will escalate

NIMS also required development of a national resource typing protocol. Under NIMS, resources are typed based upon how capable a given resource is for the task or incident. Type I resources are most capable while Type IV are least capable with Types II and III somewhere in between. Typing can help managers identify the best resources to request and assign at an incident.

NIMS also typifies incidents as an additional tool for identifying needed resources. Type I incidents are the most complex. Type V incidents are the least complex. Typically, Type I incidents require federal level resources to mitigate the emergency. Type V incidents can be mitigated using local resources.

Incident Communications

Face-to-face communications are ideal whenever possible. All responders have certain communication responsibilities at a scene:

- Briefing others when necessary
- Providing updates and evaluations of their actions
- Communicating hazards

- Acknowledging that communications have been received
- Asking for clarification whenever communications are in doubt
- Asking for more information whenever needed

Local protocols will dictate the specific details of incident communications such as radio frequencies, reporting procedures, and emergency communication. From a broad, incident management perspective, communications can be categorized as:

- Formal
- Informal
- Briefings

Formal Communications

Formal communications are those that must follow the chain of command including:

- Receiving and giving assignments
- Requesting additional resources
- Requesting support
- Reporting progress on assigned tasks
- Reporting personnel accountability
- Reporting hazards or changing conditions at the scene

Informal Communications

Informal communications include exchange of information at the scene for task completion. These are the communications between crew members as they carry out their duties. Communications between supervisors and subordinates can also be informal as long as they don't fall into formal categories. Incident safety officers may informally correct firefighter behavior as part of monitoring the scene. Emergency communications such as MAYDAYs and calls for assistance could fall into either formal or informal categories depending upon local protocols.

Briefings

Briefings are intended to pass along vital information and should not result in long conversations or complex decision making. Briefings about operational assignments, for example, should quickly and accurately describe tasks assigned, the purpose of the tasks, and the expectations for completion. Common topics for briefings include the following:

- Current state of the incident and objectives
- Safety issues, scene hazards, and emergency procedures
- Work tasks
- Work areas and facilities
- Communication protocols and reporting procedures
- Performance expectations and strategic goals
- Location resources, supplies, equipment, and the process for acquiring them

- Work schedules at longer incidents
- Opportunity to raise questions and concerns and provide feedback

NIMS-ICS generally includes three levels of briefings as follows:

- **Staff-Level:**
  — Delivered to nonoperational resources and support resources at the ICP.
  — Occur at the beginning of an incident and as necessary throughout the incident.
  — Intended to communicate the following:
    - Tasks and scope of work at the incident
    - Reporting schedule
    - Responsibilities and delegated authority
    - IC or supervisor's expectations
    - Describe the workspace, location of supplies, and work schedule

- **Field-Level:**
  — Delivered to individual, operational resources, crews, Strike Teams, or Task forces working at or near the incident site.
  — Occurs at the beginning of an operational shift.
  — Intended to communicate specific tasks, reporting relationships, and expectations.

- **Section-Level:**
  — Delivered to an entire Section; usually includes an update on the operational period during longer incidents.
  — Intended to share incident-wide direction from the IC and how those directions will affect each Section. May include:
    - Establishing Section staffing requirements
    - Assigning Section work tasks
    - Setting Section-wide scheduling rules
    - Establishing timelines for meetings and completion of tasks

- **Operational Period:**
  — Delivered at the beginning of operational period at a lengthy incident.
  — Provides similar information as other briefings with an emphasis on just the upcoming operational period.

Incident Action Plan (IAP)

The incident action plan (IAP), written or verbal, is based on information gathered during the incident size-up. The majority of emergency incident operations will be managed with a verbal IAP that is dynamic to the changing incident conditions. The IC may also use a tactical worksheet to track units and make field notes about the incident. The verbal IAP with the tactical worksheet may evolve into a written IAP as the incident grows in size and/or complexity.

Verbal IAP

The IC will communicate the incident objectives of the IAP to units and individuals who are operating at the scene and tasked with a specific work assignment. This communication is done in person or over designated radio frequencies. All incident personnel must function within the scope of the IAP. Incident personnel should direct their actions toward achieving the incident objectives, strategies, and tactics specified in the plan. When all members understand their positions, roles, and functions in the ICS, the system can safely, effectively, and efficiently use resources to accomplish the plan.

Assigned Incident Tasks

All personnel should be working towards the common goal of the IAP. Personnel operating outside the established IAP or freelancing are a danger to themselves and all other personnel on scene. A properly implemented ICS structure with clear incident objectives should be sufficient to address this concern.

NIMS IAP Planning Process

As the incident grows or has the potential for involving multiple units or agencies for an extended period, the IAP may need to be in written form. A written IAP should be forecasted early in the incident operations. By forecasting this need, the IC can expand the ICS structure to include a formalized planning process in which the written IAP is developed. Following the NIMS Planning "P" is an effective and standardized approach for developing the IAP and for all command and general staff personnel to understand their responsibilities in this process. Standardized ICS forms are available to record the various elements of the plan.

IAPs usually contain the following elements:

- **Tactical worksheet.** Basis for the development of an IAP.
- **Incident briefing.** Serves as an initial action worksheet (ICS 201 Form).
- **Incident objectives.** Objectives should be SMART: Specific, Measurable, Action-Oriented, Realistic, and Time Frame (ICS 202 Form).
- **Organization.** Description of the ICS table of organization, including the units and agencies that are involved (ICS 203 Form).
- **Assignments.** Specific unit tactical assignments divided by Branch and Division (ICS 204 Form).
- **Support materials.** Includes site plans, access or traffic plans, locations of support activities (staging, rehabilitation, logistics, and others), and similar resources.
- **Safety message.** Information concerning personnel safety at the incident; may also be part of the Incident Safety Plan that the Incident Safety Officer develops (ICS 208 or 208H Form).

The written IAP is maintained at the Incident Command Post and updated or revised as warranted or at the end of the specified time interval (NIMS *Planning "P"*). At the end of the incident, the plan is used as part of the postincident analysis and critique.

Key Terms

Chain of Command — Order of rank and authority in the fire and emergency services.

Command Staff — The command staff consists of the Information Officer, Safety Officer and Liaison Officer. They report directly to the Incident Commander and may have an assistant or assistants, as needed. (National Wildfire Coordinating Group (NWCG) *Glossary of Wildland Fire Terminology*).

Incident Action Plan (IAP) — Contains objectives reflecting the overall incident strategy and specific tactical actions and supporting information for the next operational period. The plan may be oral or written. When written, the plan may have a number of attachments, including: incident objectives, organization assignment list, division assignment, incident radio communication plan, medical plan, traffic plan, safety plan, and incident map. *Formerly called* shift plan. (National Wildfire Coordinating Group (NWCG) *Glossary of Wildland Fire Terminology*).

Incident Commander (IC) — Person in charge of the Incident Command System and responsible for the management of all incident operations during an emergency.

Incident Command System (ICS) — Standardized approach to incident management that facilitates interaction between cooperating agencies; adaptable to incidents of any size or type.

Memorandum of Understanding (MOU) — Form of written agreement created by a coalition to make sure that each member is aware of the importance of his or her participation and cooperation.

Span of Control — Maximum number of subordinates that that one individual can effectively supervise; ranges from three to seven individuals or functions, with five generally established as optimum.

Staging Area — Locations set up at an incident where resources can be placed while awaiting a tactical assignment on a three (3) minute available basis. Staging Areas are managed by the Operations Section. (National Wildfire Coordinating Group (NWCG) *Glossary of Wildland Fire Terminology*).

Unified Command — In ICS, unified command is a unified team effort which allows all agencies with jurisdictional responsibility for the incident, either geographical or functional, to manage an incident by establishing a common set of incident objectives and strategies. This is accomplished without losing or abdicating authority, responsibility, or accountability. (National Wildfire Coordinating Group (NWCG) *Glossary of Wildland Fire Terminology*).

References

National Wildfire Coordinating Group (NWCG) *Glossary of Wildland Fire Terminology*. Accessed online. https://www.nwcg.gov/glossary/a-z

Chapter 4

Strategy and Tactics

Table of Contents

Parts of a Ground Cover Fire 81
- Origin 82
- Head 82
- Fingers 82
- Perimeter 82
- Heel 82
- Flanks 82
- Spot Fires 82
- Islands 82
- Slopover 83
- Green 83
- Black 83

Size-Up 83

Forming an Incident Action Plan (IAP) 85

Fire Control Strategies 86
- Direct Attack 88
- Indirect Attack 89
- Locating and Developing the Fireline 90
- Anchor Points 91
  - *Fireline Width* 91
  - *Fireline Construction* 92
  - *Line Construciton with Mechanized Equipment* 96

Key Terms 97

References 98

Strategy and Tactics

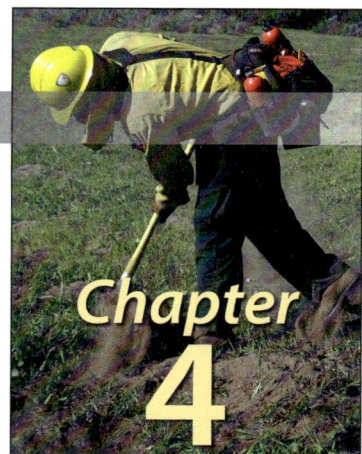

Chapter 4

Firefighters use a variety of methods to control and extinguish ground cover fires. The most appropriate methods provide for personnel safety and make the best use of the available resources. This chapter describes the parts of a ground cover fire as well as size-up, forming an Incident Action Plan (IAP), and fire control strategies.

Parts of a Ground Cover Fire

Firefighters must use standard names for various parts of a ground cover fire to provide a clear report on conditions, to give direction, and to provide for firefighter safety. The parts of a ground cover fire are named for their unique characteristics and locations. The names used to identify the parts of a typical ground cover fire are shown in **(Figure 4.1)**. Every ground cover fire contains at least two or more of the following parts:

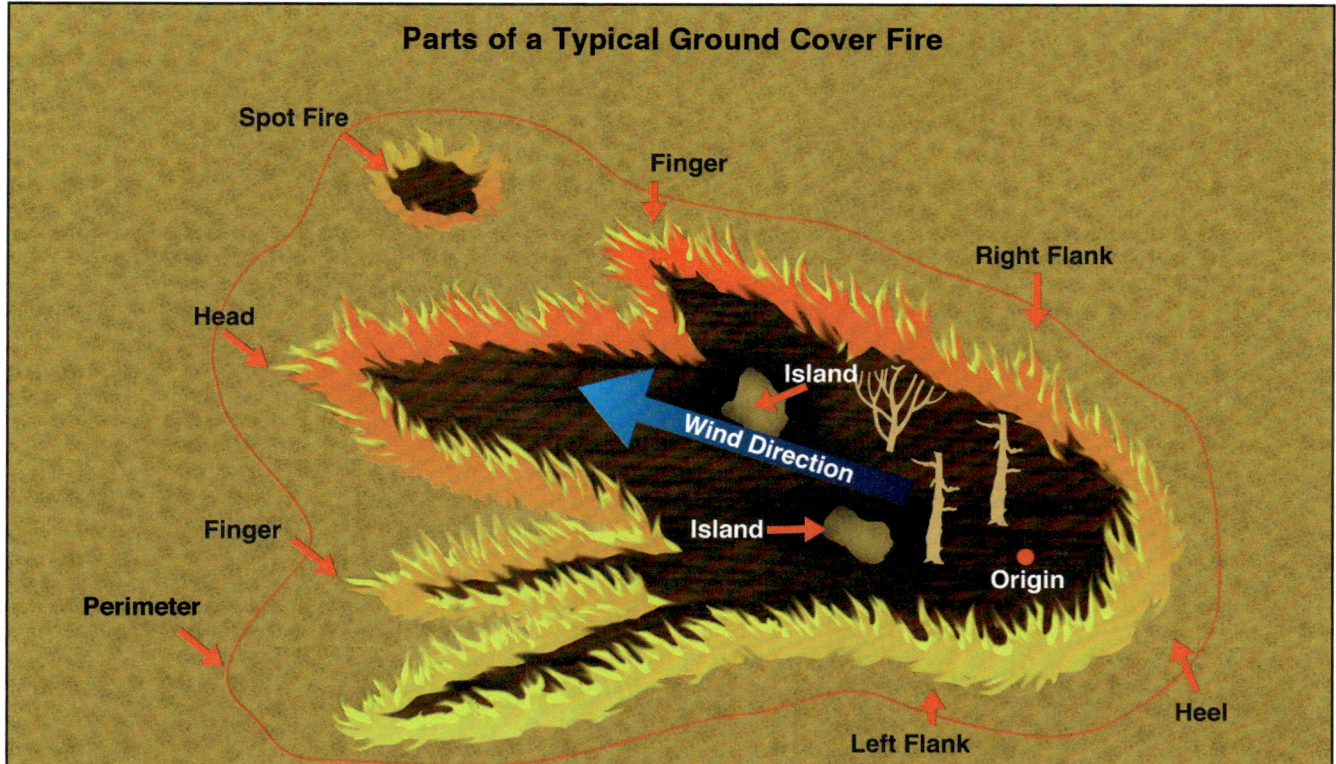

Figure 4.1 All ground cover fires share two or more of a set of common parts.

Origin

The **origin** is the area where the fire started and the point from which it spreads. It is dependent on the availability of fuel and the effects of wind and slope. The origin is often next to a trail, road, or highway, but it also may be in very inaccessible areas, such as those where the fire is started by lightning or campfires. The area of origin should be protected for subsequent investigation of fire cause whenever possible.

Head

The **head** is the part of a ground cover fire that spreads most rapidly. The head is usually found on the opposite side of the fire from the area of origin and in the direction toward which the wind is blowing. The head burns intensely and usually does the most damage. Usually, the key to controlling the fire is to control the head and prevent the formation of a new head.

Fingers

Fingers are long, narrow strips of fire extending from the main fire. They usually occur when the fire burns into an area that has light fuel and patches of heavy fuel. Light fuel burns faster than the heavy fuel, which gives the finger effect. When not controlled, these fingers can form new heads.

Perimeter

The **perimeter** is the outer boundary, or the distance around the outside edge, of the burning or burned area. It will continue to grow until the fire is suppressed. Also commonly called the **fire edge**.

Heel

The **heel**, or *rear*, of a ground cover fire is the end opposite the head. Because the heel usually burns downhill or against the wind, it burns slowly and quietly and is easier to control than the head.

Flanks

The **flanks** are the sides of a ground cover fire, roughly parallel to the main direction of fire spread. The right and left flanks separate the head from the heel. It is from these flanks that fingers can form. A shift in wind direction can change a flank into a head.

Spot Fires

Flying sparks or embers landing outside the main fire cause **spot fires**. Spot fires present a hazard to personnel (and equipment) working on the main fire because they could become trapped between the two fires. Spot fires must be extinguished quickly or they will form a new head and continue to grow in size.

Islands

Patches of unburned fuel inside the fire perimeter are called are called **islands**. Because they are unburned potential fuels for more fire, they must be patrolled frequently and checked for any spot fires.

Slopover

Slopover occurs when fire crosses a control line or natural barrier intended to confine the fire (**Figure 4.2**). Slopovers differ from spot fires mainly in their location relative to the control line. A slopover occurs immediately across and adjacent to the control line; a spot fire occurs some distance from it.

Figure 4.2 A firefighter mops up a slopover.

Green

The area of unburned fuels next to the involved area is called the **green**. The green area of a ground cover fire should not be confused with the *green zone* often indicated at hazardous materials incidents or fire scenes. Green at ground cover fires is simply the opposite of the burned area (the black) and does not indicate that the area is safe (**Figure 4.3**).

Figure 4.3 The green.

Black

The opposite of the green, the **black** is the area in which the fire has consumed or "blackened" the fuels. The black can be a relatively safe area during a fire but is often hot and smoky with numerous hot spots and smoldering snags (standing dead trees), stumps, and downed trees (**Figure 4.4**). However, the black may not be safe in steep terrain, where flames from adjacent, unburned fuels can project intense radiant heat into it.

Figure 4.4 The black.

Size-Up

Size-up is an ongoing process of evaluating current and expected fire conditions. It starts before a call comes in from the dispatcher and continues throughout an incident. All involved fire personnel should constantly monitor fire conditions and make adjustments as needed. Size-up involves developing a mental picture of current, expected, and potential behavior of a fire. It also involves evaluating the threat to life, property, and resources. The initial size-up consists of gathering information prior to and upon arrival at scene. To make an initial size-up, the information can be broken down as follows:

Before Dispatch

- Preincident plans
- Fire weather forecasts
- Current/predicted fire danger indices and conditions
- Local terrain and fuels
- Fire behavior
- Ground cover resources available (including backup)

> **Size-Up En Route**
> - Smoke (volume, color, and movement)
> - Clouds (type, size, and movement)
> - Time of day
> - Weather (temperature, humidity, and wind)
> - Best access (route in)
> - Jurisdiction (government owned or private)
> - People leaving the area (vehicles, license numbers, etc.)
> - Communications (command, tactical channels assigned while en route)

> **Size-Up on Arrival**
> - What is burning (fire behavior, rate of spread, size, etc.)
> - What is threatened (structures, timber, etc.)
> - Topography
> - Populations at risk
> - Access to the fire (vehicles, air, or personnel)
> - Initial safety zones/escape routes
> - Special hazards (snags, hazardous materials, downed wires, etc.)
> - Point of initial attack
> - Most likely area of origin
> - Resources needed

The information gathered prior to and upon arrival helps the initial Incident Commander make one of the first and most important decisions of an incident — whether the types and numbers of resources at the scene or en route are sufficient to control this fire and if additional resources need to be called. If the initial size-up indicates that additional resources or different types of resources are needed, they must be ordered as soon as possible. If unsure of what types and numbers of resources are needed, the IC should order any that *might* be needed. Units that prove unnecessary can be held in staging or returned to quarters.

Once on scene, the IC must focus on the situation at hand: where and how to invest the available resources to do the best until other resources arrive. To do this, the IC should answer the following questions:

- What is the most important work to be done first?
- Where can the most effective work be done?

One of the simplest ways to make these decisions is to meet the three incident priorities: life safety, incident stabilization, and property conservation.

The IC transmits a clear and concise *report on conditions* for the other incoming units. The report describes:

1. What has happened (fire history): *"Fire is 10 acres in size."*
2. What is happening (current size-up): *"Fire is burning up O'Farrell Hill."*
3. What is going to happen (fire behavior prediction without suppression action, initial objectives, and actions): *"Attacking right flank to protect subdivision."*

The report confirms the address or location of the fire and gives the best means of access to it. For example: "Ten acres involved, moderate rate of spread up O'Farrell Hill, attacking right flank to protect subdivision. All units enter off Jones Road."

The IC may decide that the resources at scene are sufficient to handle the situation and return other responding units. If not, the IC may begin to expand the Incident Command Organization, initiate those actions within the capabilities of at-scene units, and assign objectives to those still en route. For example:

- "Engine 7182 is O'Farrell Command.
- Engines 7188 and 5660 take the right flank.
- Engine 8141 take the left flank."

Forming an Incident Action Plan (IAP)

The purpose behind following good size-up protocols is to develop an Incident Action Plan (IAP) (whether written or verbal) and to prepare to meet the requirements to initiate that plan. One of the simplest ways to formulate the plan and to meet the three (3) incident priorities — life safety, incident stabilization, and property conservation — is to organize in your mind or on paper your management objectives. Managing by objectives gives you a way to measure/evaluate whether or not your plan is working or if adjustments need to be made. This technique also gives the responders a clear operating picture of what needs to be accomplished and a picture of whether or not you have enough resources on hand or en route to accomplish the mission.

> **SMART**
>
> When establishing your objectives, note them in the SMART format:
>
> **S**pecific
>
> **M**easurable
>
> **A**chievable
>
> **R**ealistic
>
> **T**ime Sensitive
>
> For example:
>
> - Provide for the safety of the responders and public.
> - Contain the fire south of O'Farrell Hill, west of Kings Orchard, north of Skull creek, and east of Jones road.
> - Evacuate the ranch houses along O'Farrell Hill by 15:30.

The plan should also provide for the following:

- Strategy (broad-scope idea) on how to accomplish the mission. An example of this may read or sound like this: *"Utilizing a combination of offensive and defensive suppression modes contain the fire east of Jones Road."*
- Protection of any evidence at the point of origin.
- Resource capabilities match current and expected fire behavior
- Flexibility for responding to changing fire conditions.
- Outline of tactics (specific hands-on tasks) to be initiated to accomplish the strategy.

In an extended attack operation, the Operations Section Chief would outline the tactics to be utilized to accomplish the strategy. An example of this may read or sound like this: *"A firing group, with support from brush trucks, will conduct burn out operations along Jones Road."* Unlike structure fire fighting protocols, where the offensive and defensive modes are rarely combined, they are often combined when fighting ground cover fires.

After the plan has been developed, the IC in the initial attack scenario or the Operations Section Chief in the extended operations situation would deploy the resources in one of three modes:

- *Offensive*
- *Defensive*
- *Combination*

This decision most often depends on the kind, type, and amount of resources on scene and available and the size and behavior of the fire. Also taken into account are the fuel and topographic elements that may inhibit the use of certain types of resources. If ample and appropriate resources are available, an offensive mode can be used to accomplish the mission. Fire behavior, intensity, and the kind and type of resources available will determine if the offensive actions will be direct or indirect. If resources are limited, then a defensive mode may be employed in order to accomplish some of the incident objectives until additional resources arrive or fire behavior changes.

Fire Control Strategies

The *direct* and *indirect* approaches are the two most basic methods for attacking ground cover fires. A **direct attack** is action taken directly against the flames at its edge or closely parallel to it. The **indirect attack** is used at varying distances from the advancing fire **(Figure 4.5)**. This method is generally used against fires that are too hot, too fast, or too big for a direct attack. As with any other type fire, size-up must continue throughout a ground cover fire so that these adjustments can be made when required (NWCG).

Figure 4.5 Firefighters making a direct attack on a ground cover fire (left) and an indirect attack (right).

Either type of attack may also be referred to as either flank attacks or **parallel attacks**. Flank attacks are normally used for moderately intense fires and can be either direct or indirect. The attack begins at a secure **anchor point** (road, highway, body of water, previous burn) on one or both of the fire's flanks and works toward the head. A parallel attack is a form of indirect attack. A parallel attack involves creating a **control line** quickly using dozers or other heavy equipment. The control line can be constructed much closer to the fire than a line created with hand tools, and more quickly **(Figure 4.6)**. This attack sacrifices less territory to the fire as it burns and frees up firefighters to patrol the control line.

The methods used to attack ground cover fires revolve around perimeter control. The control line may be established at the burning edge of the fire, next to it, or at a considerable distance away. The intent of any well-developed plan is to start work on suppressing the fire as soon as possible. In initiating fire attack operations, determine the following:

- Primary/secondary access
- Location of escape routes
- Special hazards such as burning snags, hazardous materials, etc.
- Good anchor points such as roads, burned area, etc.
- Where to attack fire (head or flank)
- How to attack fire (direct or indirect)
- Type of control line needed (wet lines, scratch lines, dozer lines, etc.)
- Existing barriers that can be used
- When next units will arrive
- How topography will affect fire behavior
- Most likely point of origin

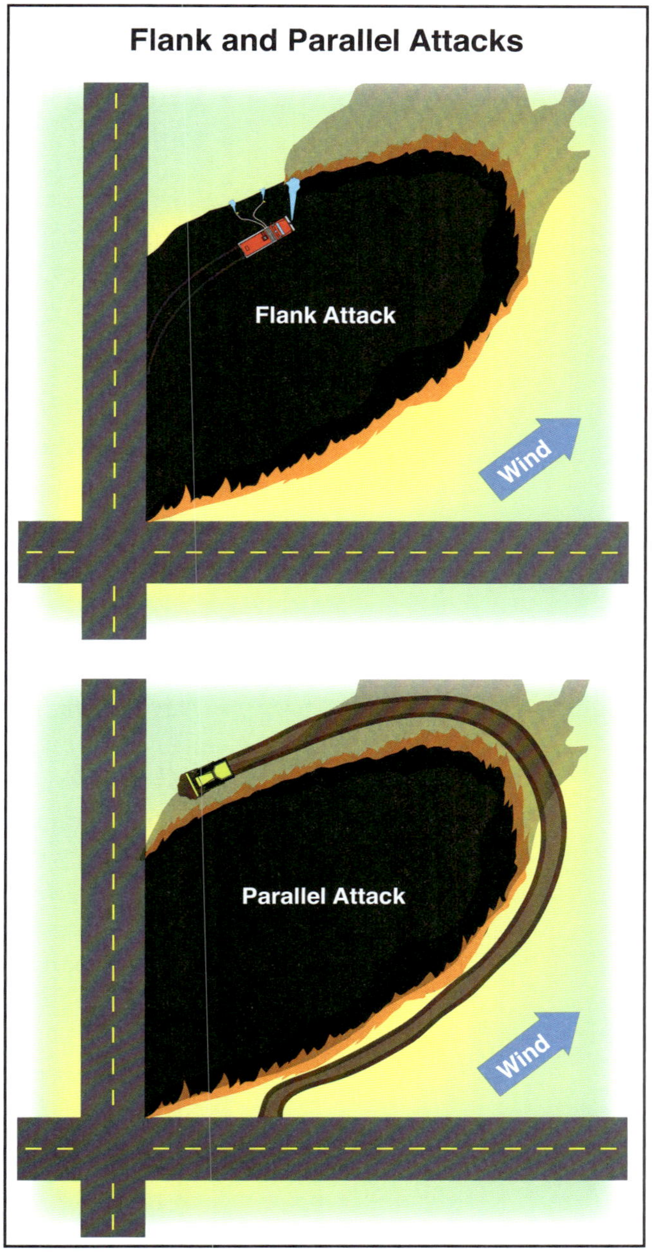

If the initial attack plan is working, follow the plan. If not, adjust the plan and implement the following changes:

- Notify the dispatch/communications center immediately if a fire exceeds the capabilities of the at-scene forces.
- Anticipate the need early for additional resources and stage them so that they are available if the situation deteriorates.
- Inform the dispatch/communications center of the progress on the fire and of any significant changes.

Figure 4.6 Illustrating flank and parallel attacks for ground cover fires.

At the earliest opportunity, transmit the following information to the dispatcher:

- Designated IC
- Fire name
- Location
- Access
- Staging area locations
- Terrain
- Size of fire
- Evacuations needed for residents, campers, etc.
- Anticipated control problems (status/progress update)
- Values threatened
- Anticipated time of control
- Weather
- Resources needed, if any
- Fire behavior
- Any known hazards in the area

Direct Attack

As the name implies, direct attack suppresses a fire by attacking the flame front directly from within or from outside the burned area. A direct attack may involve extinguishing the fire by cooling with water or foam or interrupting the flaming process with chemicals. While the use of water makes the attack more efficient, in many cases a direct attack involves smothering the fire with dirt or removing the fuel (**Figure 4.7**).

Figure 4.7 A firefighter throws dirt onto a fire.

A direct attack is an aggressive, offensive attack at the edge of the fire. It is normally used on relatively small fires (flame lengths of no more than 4 feet [1.2 m]) where heat and smoke do not keep firefighters from working at the fire's edge. Like every other tactical decision, the decision to order a direct attack must weigh the risks against the potential benefits. Whenever conditions permit its use, direct attack is often the strategy of choice for:

- Small fires or larger fires of low-to-medium intensity
- Hotspotting (knocking down flare-ups [hot spots] ahead of line construction or other suppression activities.
- Extinguishing fingers spreading from the main fire.
- Running (rapidly spreading) fires in light fuels.
- Fires where the values at risk (life safety, private property and public infrastructure) must be kept to a minimum.

The primary advantage of a direct attack is that firefighters are close to or working in the burned area, which may be used as a safety zone in the event of threatening conditions.

Major disadvantages of using a direct attack:

- Firefighters are exposed to heat, smoke, and flame during operations whether utilizing hoselines or constructing a fireline directly adjacent to the edge of the fire.
- Embers may blow across the line and start spot fires.
- Constructing firelines with hand tools is physically taxing to firefighters.

Direct attack considerations to remember:

- Conduct control efforts, including line construction, at the fire perimeter that becomes the control line.
- Use when the fire perimeter is burning at low intensity, permitting safe operation at the fire's edge.
- Implement where high-value resources and/or improvements are threatened.
- Use when the amount of area burned is kept to a minimum.

Indirect Attack

Indirect attack sacrifices a certain amount of vegetation or other exposed values. Indirect attack is used when:

- The intensity of a fire makes direct attack unsafe.
- The fire develops long fingers.
- There is not enough time to develop a control line at a fire's edge.

In an indirect attack, fire suppression forces are withdrawn to roads, trails, or other natural fuel breaks or to a preconstructed control line. Personnel must identify existing safety zones or construct new ones.

The fuel between these barriers and the fire is burned out or backfired. While indirect attack does not force firefighters to work as close to a fire front as does direct attack, an indirect attack is not without risk. Firefighters can be in jeopardy during indirect attacks for the following reasons:

- While the control line is being constructed, the fire is growing in size and intensity.
- Firefighters are often unable to directly observe the behavior of the fire because the line is constructed some distance from the fire's edge. This increases the need for Lookouts, Communications, Escape Routes, and Safety Zones (LCES). See Chapter 2 for additional information on LCES.
- It is always risky to have firefighters work in an area with unburned fuel between them and the main fire.
- Whenever firefighters are engaged in indirect suppression operations, they must be especially mindful of the 18 "Watchout!" Situations listed in Chapter 2.

Disadvantages of an indirect attack include the following:

- The fuel left inside a control line can allow a fire to increase and reach such intensity and rate of spread that it could jump the control line.
- The unburned fuel between the main fire and the control line can also allow the fire to develop a large convection column, increasing the risk of spot fires. This prospect is increased if the winds suddenly pick up after the attack is underway.
- Changes in wind direction, fuel types, and topography could also change the direction of fire spread, rendering the control line ineffective. These dangers reinforce the importance of continually sizing up and adapting to changing conditions.
- More fire entrapments and fatalities occur on indirect attack than on direct attack, so indirect attack is generally considered more dangerous.

In summary, the following facts about indirect attack:

- Indirect attack is located on the control line along man-made or natural firebreaks, along favorable breaks in topography, or at considerable distances from the fire. The intervening fuel is burned out.
- If indirect attack is necessary, the fire may be rapidly growing to the point of requiring extended attack (needing more resources).
- Indirect attack is used on crown fires; steep terrain; fast-moving ground fires too intense for engine, brush trucks, or hand crews; or in areas with constructed or natural barriers.

Locating and Developing the Fireline

One of the IC's most important initial decisions concerns the location of the fireline in relation to the fire's edge. The location of the control line is related to the method of attack, offensive or defensive. Base the decision of where to locate the fireline on the following considerations:

- Ensure that the safety of personnel and equipment can be established and that LCES can be started.
- Make sure that the type, kind, and experience of the resources are establishing the line.
- Consider where it is impractical or unsafe to initiate a direct attack.
- Locate the line far enough from a fire to be completed, burned out, and

held before the fire reaches it with predicted rate of spread and fire behavior.
- Allow enough time to permit forces not only to build lines but also to do other needed work, such as felling snags, if necessary.
- Make the line as short and straight as practical.
- Select the easiest routes for control as long as you do not compromise line effectiveness. Excessive area or values are not sacrificed.
- Eliminate possible hazards from the fire area, and provide a safe distance between lines and hazards that must be left in the fire area.
- Use mechanized equipment for line construction where possible.
- Encircle areas where spot fires are numerous and the unburned fuels safely burned out.
- Avoid undercut/underslung lines (lines below a fire on a slope) and sharp turns in the line.
- Never construct the fireline downhill with fire below you and there is unburned fuel between you and the fire.
- Minimize environmental effects and follow agency policy.

Anchor Points

All control lines must have an *anchor point* — a barrier to fire spread where the control action begins. The purpose of an anchor point is to prevent a fire from burning around the end of the control line and possibly outflanking the suppression crews and placing them in danger. An anchor point is usually not a constructed fireline, but it could be one of the following:

- Hose lay
- Secured aerial retardant drop
- Road
- Bare field
- Stream
- Cliff
- Previously burned section of the fire

> **CAUTION:**
> **Starting fireline construction from an anchor point is critical to firefighter safety.**

Fireline Width

Weather, topography, and the arrangement, volume, and type of fuels all combine to dictate the width of line needed. In sparse surface fuels, such as duff or light grass, the line may only need to be a foot or two wide. In heavier fuels and in severe burning conditions, the line must be wider. The fuels between the control line and the fire's edge should be burned out, particularly in medium to heavy fuels, to effectively widen the control line. When a fireline must be constructed by hand crews only, it is important to save time and conserve the firefighters' energy by making the line only as wide as necessary. Most firelines vary in width from a foot to a few yards; in general, the hotter and faster a fire burns, the wider the line must be. Anything that affects how a fire burns must be considered when deciding how wide the fireline must be.

One guideline to determine line width is to figure at least one and one-half times the height of the fuels carrying the fire or two and one-half times the height of the flames. When backfiring in areas subject to extreme fire behavior, such as on steep slopes, fireline width should be at least two times the fuel height. A supervisor or experienced firefighter should make the decision regarding line width in any specific situation based upon the tactical determination of direct or indirect methods. The most important factors in determining fireline width are fuel, slope, weather, part of fire, and fire intensity.

Fireline Construction

Firelines are constructed by using hand tools or mechanized equipment, such as dozers, motor graders or tractor-plows, to remove surface and subsurface fuels down to mineral soil **(Figure 4.8)**. In this process, break the continuity of any aerial fuels over the line to make them unavailable to a fire. Break up and disperse concentrations of surface fuels close to the control line. A pile of brush located inside but adjacent to the control line is a potential source of a hot spot, flare-up, or slopover. The same type of pile located outside the line serves as a bed of fuel that is available to radiant heat or airborne sparks or embers from the fire which could start a rapidly developing spot fire in this pile. Cut down snags and other standing aerial fuels near the line. If time allows and they are unburned, move them outside of the control line.

Figure 4.8 Dozers construct firelines to remove surface fuels down to mineral soil.

Trench a control line built on a slope below a fire to stop materials, such as logs, pine cones, or other potentially hot material, from rolling into unburned fuels and spreading the fire. A good trench has an earthen berm to stop burning materials. To stop a fire that is burning upslope, construct a control line just over the ridge on the other side **(Figure 4.9)**. A line in this location uses the effect of slope on fire behavior to decrease the possibility of the fire jumping the line. Even though heavy equipment can construct a fireline much faster than hand crews, the use of hand tools for fireline construction is quite common. In many situations, such as in remote, rugged, or steep terrain or during certain seasons such as springtime (when the ground may be soft but the surface fuels dry enough to burn), using engines, brush trucks, or heavy equipment is impractical, unsafe, or prohibited by environmental protection regulations. In these situations, firefighters with hand tools may be the only practical means of constructing a fireline.

Figure 4.9 A wide fireline is needed above fires on steep slopes.

When constructing firelines, crew members typically walk and work 10 feet (3 m) apart for safety. This spacing helps prevent crew members from being struck by the handles or heads of tools while they carry them or use them in constructing a line. Firefighters should give a loud verbal warning such as "tool coming through!" when they need to pass close to each other on a line, especially if visibility is reduced by smoke or darkness.

Several methods are used to organize hand crews for constructing a fireline. One common method (called *leapfrogging*) uses the following procedures:

- Assigns each crew member a few feet (meters) of the line.
- Completes that portion of the line before moving to another portion **(Figure 4.10, p. 94)**.

Another common method known as *progressive line construction* (also called the *one-lick method*), uses the following procedures:

- The crew is arranged in a staggered line, and each member remains in position relative to the other members as the line construction progresses **(Figure 4.11, p. 94)**.

Strategy and Tactics • Chapter 4   93

Figure 4.10 Each firefighter is responsible for a portion of the fireline.

Figure 4.11 Firefighters cut a progressive fireline.

- Each member takes one stroke ("lick") with the tool before moving one step forward to repeat the action.
- The crew works in unison until the line is completed.

This method of line construction requires teamwork but promotes safety and efficiency because no one passes anyone else on the line.

Regardless of how a crew is organized, they most often work along the line in what is called a *typical tool order*. For example:

- The order is dictated by the type of fuel to be cleared.
- In light fuels, such as pasture land or mowed field crops, smothering, raking and scraping tools such as Swatters, Council Rakes and Mcleods, may need to be used.

In medium fuels, the typical tool order might be:

- Chain saws, brush hooks, or Sandvigs
- Pulaskis or Rouge Hoes
- Shovels or McLeods

In heavier fuels, members with chain saws, axes or brush hooks lead the crew followed by Pulaskis, Rouge Hoes, McLeods, Combi-tools, and shovels.

Safe and efficient line construction by hand crews depends on having the right tools for the job and using those tools properly. Some tools are more effective in certain fuel types than others, and knowledge of the fuel types and the topography at the fire helps in selecting the right tools. When tools are issued, crew members should inspect them to be sure they are sharp and in good condition.

In situations where the fuel through which the line is to be cut is primarily dry leaf litter and duff, some agencies are using leaf blowers to construct a line. The light, dry material is blown away to expose the bare soil beneath. Other firefighters carrying cutting and scraping tools may have to follow the blower operator to clear small bushes and clumps of grass from the line. Chain saws can make line construction much easier, but they must be used with care. Only trained and experienced personnel wearing protective eye wear, hearing protection, gloves, and protective chaps should operate chain saws **(Figure 4.12)**.

Figure 4.12 Only personnel wearing proper PPE should operate chain saws.

Do not transport chain saws with the motor running. When not in use, cover the chain with a safety guard. Whenever possible, turn off saws and allow them to cool before being refueled. To make a line as effective as possible, the following must be done:

- Remove all vegetation and debris from the line.
- Clear the line down to mineral soil.
- Widen the line enough to safely burn out from it.
- Throw all burned/charred material into the black.
- Scatter all cut and unburned fuels into the green (unless needed for burning out).
- Remove all branches that hang over the line.

Line Construction with Mechanized Equipment

If it is available and if the topography allows for its safe and effective operation, using mechanized equipment is a fast and efficient way to build a control line. Various kinds of mechanized equipment, such as dozers, graders (road maintainers), and tractor-plows, are used for constructing lines. In an emergency, farm equipment, such as farm tractors with plows or discs, can be effective in situations where they can operate safely. When conditions allow, mechanized equipment can build a wide control line much faster than hand crews. Dozers can cut a one-blade-width control line at about one-half mile or 880 yards (805 m) per hour, subject to several variables. In general, the newer and larger the dozer, the faster it can construct a line.

The rate of construction is affected by the following:

- Degree of slope
- Presence or absence of rocks
- Fuel type
- Direction of line construction
- Operator skill
- Operator fatigue
- Availability of lights for night work
- Atmospheric temperature

In areas where ground cover fires occur frequently, mechanized equipment specifically designed for building control lines is commonly used. Crawler tractors or all-wheel-drive vehicles with fire plows can build lines quickly and easily if the topography and soil conditions are right for their use. Dozers and other mechanized equipment are typically brought to a scene on low-boy or tilt-bed transports **(Figure 4.13)**. Transporting this equipment to a point close to a line may present problems because of the length and weight of the transport vehicles. In addition, unless dozer tenders or other support units accompany the mechanized equipment into remote locations, other provisions must be made to refuel and service the equipment. This may require that fuel and lubricants be flown into the area by helicopter.

Operating heavy equipment on ground cover fires requires all of the normal safety procedures for using mechanized equipment and the additional safety

Figure 4.13 Dozers and other mechanized equipment are typically brought to a scene on low-boy or tilt-bed transports.

procedures related to fire suppression with hand crews. Even routine preventive maintenance becomes a safety issue if a mechanical breakdown might result in a unit being overrun by a fire. One of the most important safety practices around mechanized equipment is maintaining effective communication between the equipment operator and those on the ground. Using a lookout (swamper) is a common way of meeting this requirement. Because it is possible for dozers and other equipment to be overrun by a blowup, another important safety feature for the equipment operators is the addition of a fire shelter to their personal protective equipment. (For more information on fire shelters, see Chapter 2.) In some areas, completely enclosed and air-conditioned environmental cabs are used to protect the operator.

Operation of dozers or other heavy equipment is usually safer and more efficient if they are used in pairs. The operator of the lead dozer can rough out a control line and remove some of the fuel. The operator of the second machine can complete the line construction by removing the rest of the fuel down to mineral soil. Operators can also help each other if either machine becomes stuck or stalled. By working together, they can quickly build a safety zone if they are in danger of being overrun by a fire.

> **WARNING**
> **Dozers working on slopes frequently dislodge rocks. Therefore, firefighters should never work downslope from mechanized equipment without posting a lookout.**

Key Terms

Anchor Point — An advantageous location, usually a barrier to fire spread, from which to start constructing a fireline. The anchor point is used to minimize the chance of being flanked by the fire while the line is being constructed. (National Wildfire Coordinating Group (NWCG) *Glossary of Wildland Fire Terminology*).

Black — Area already burned by a ground cover fire. *Also called* Burn.

Control Line — Inclusive term for all constructed or natural barriers and treated fire edges used to control a fire. (National Wildfire Coordinating Group (NWCG) *Glossary of Wildland Fire Terminology*).

Direct Attack — To attack a ground cover fire directly at the burning edge.

Fingers of a Fire — The long narrow extensions of a fire projecting from the main body. (National Wildfire Coordinating Group (NWCG) *Glossary of Wildland Fire Terminology*).

Fire Edge — The boundary of the burned or burning material at any given time.

Flanks of a Fire — Parts of a fire's perimeter that are roughly parallel to the main direction of spread. (National Wildfire Coordinating Group (NWCG) *Glossary of Wildland Fire Terminology*).

Green — Area of unburned fuels, not necessarily green in color, adjacent to but not involved in a ground cover fire.

Head of a Fire — The most rapidly spreading portion of a fire's perimeter, usually to the leeward or up slope. (National Wildfire Coordinating Group (NWCG) *Glossary of Wildland Fire Terminology*).

Heel — Rear portion of a ground cover fire. *Also called* Rear.

Indirect Attack — Controlling the fire by locating the control line along natural firebreaks some distance from the approaching fire and burning out the intervening fuels.

Island — Unburned area within a fire perimeter.

Origin — Point of original ignition of a fire.

Parallel Attack — Method of fire suppression in which fireline is constructed approximately parallel to, and just far enough from the fire edge to enable workers and equipment to work effectively, though the fireline may be shortened by cutting across unburned fingers. The intervening strip of unburned fuel is normally burned out as the control line proceeds but may be allowed to burn out unassisted where this occurs without undue delay or threat to the fireline. (National Wildfire Coordinating Group (NWCG) *Glossary of Wildland Fire Terminology*).

Perimeter — Entire outer edge or boundary of a fire.

Size-Up — Ongoing process of observation and evaluation of existing factors that are used to develop objectives, strategy, and tactics for fire suppression.

Slopover — A fire edge that crosses a control line or natural barrier intended to confine the fire. (National Wildfire Coordinating Group (NWCG) *Glossary of Wildland Fire Terminology*).

Spot Fire — Fires starting outside the perimeter of a main fire typically caused by flying sparks or embers.

References

National Wildfire Coordinating Group (NWCG) *Glossary of Wildland Fire Terminology*. Accessed online. https://www.nwcg.gov/glossary/a-z

Chapter 5

Ground Cover Engine Operations

Table of Contents

Structural Apparatus Used on Ground Cover Fires 103

Resource Typing-Incident Command System (NIMS-ICS) 105
    Resource Typing 106
    Wildland Engine Types 107
        Type 3 .... *107*
        Type 4 .... *107*
        Type 5 .... *108*
        Type 6 .... *108*
        Type 7 .... *108*

Fire Control Tactics 108
    Flank Attack 109
    Pincer Attack 109
    Frontal Attack 110
    Mobile Attack 110
    Tandem Attack 112
        Two Engines *112*
        Hand Crews *113*
    Hotspotting 113
    Indirect Attack 114
        Advantages of an Indirect Attack *115*
        Disadvantages of an Indirect Attack *115*

Hose Lays 115
    Booster Line 116
    Progressive Hose Lay 117

Ground Cover/Urban Interface Operations 118
    Command, Control, and Accountability 119
        Strike Team *120*
        Task Force *120*
    Ingress and Egress 120
        Residents and the Public *121*
        Evacuation *121*
        Routing Traffic and Establishing Access *121*
    Structure Triage 122
        Greatest Potential Threat *122*
        Probable Threat *122*
        Factors Affecting Triage *123*
        Consider All Factors *127*
        When Structures Cannot Be Saved *127*

Structure and Site Preparation 128
    Structure Protection: Lessons Learned 128
    The Structure 129
    On-Site Resources 129
        Locate Water Sources *129*
        Adjacent Resources *129*
    Clearance Around Structures 130
    Removing and Trimming Fuels 130

Fireline Construction 130
    Intermediate Fuels 131
    Yard Accumulation 131
    Flammable and Explosive Hazards 132

Structural Exterior and Interior Preparations 132

Private Vehicles 133

Pets and Livestock 134

Pretreatment of Structures 134
    Sprinkler Systems 134
    Class A Foam 134
    Fire Gel 134
    Structure Wrap 134

Structure Protection Tactics 134

Working Hoselines 135

Nozzles 137

Confronting the Fire at the Structure 137
    Spotting Zone 137
    Full Containment..138
    Partial Containment 138
    No Containment Possible...................................138
    Fighting Roof Fires 138

Water and Foam Use................................ 139
    Water Supply 139
        Water Application *139*
        Wetting Down with Water *139*
    Reducing the Heat Buildup 140
        Duration of the Heat Wave *140*
        Peak Heat Wave Tactics *140*
    Class A Foam 140
        Properties of Foam.. *141*
        Types of Foam *141*
    Structure Treatment 141

Hit and Run Tactics 142
    Retreating and Returning 142
    Extinguishment and Follow Up 143

Firing Operations.. 143
    Burning Out...143
    Backfiring 143
        When to Burn Out or Backfire........................... *143*
        Timing and Coordination................................. *144*
        Control Lines for Firing Operations................. *144*
    Firing and Holding..145

Follow-Up 146
    Before Leaving the Area 146
    Patrol Duties...147

Public Relations 147
    Dealing with the Media 148
    Dealing with the Public..149

Key Terms 149

References 151

Ground Cover Engine Operations

Chapter 5

Fighting a major fire of any kind would be almost impossible without modern fire fighting apparatus. This is certainly true of ground cover fires. Modern fire apparatus helps firefighters control ground cover fires quickly, effectively, and safely — thus reducing loss to life, property, and valuable natural resources. In this manual, *fire apparatus* refers to those vehicles used for fire suppression.

Structural Apparatus Used on Ground Cover Fires

The main purpose of structural fire engines is to provide adequate fire streams for attacking structure fires and other types of fires commonly encountered by municipal fire departments. However, structural fire engines can be very effective on those many fires that burn at the roadside or on a highway median strip **(Figure 5.1)**. They can be used in a quick attack role on small nuisance fires by pumping from their tanks and using small handlines. From the relative safety of a paved street or road, structural engines can also be very effective on large roadside fires by being the anchor point for *progressive hose lays*.

Figure 5.1 Structural fire engines can be very effective on fires that burn at the roadside.

> **NOTE:** Structural engines were not designed for and should never be deployed in an off-road capacity.

A progressive hose lay consists of extending hose from the apparatus to the fire's edge and extinguishing fire as the hose is extended, connecting another section, advancing, and extinguishing more fire **(Figure 5.2)**. A progressive hose lay is limited only by the friction loss in the hose and the amount of water available to the engine. However, once a reliable water supply has been established (for example, water tender shuttle), an engine can supply a large volume of water for one or more progressive hose lays at considerable distances from a road or street. Given a relatively short response distance, an adequate water supply, and trained personnel, a structural engine used in this way can keep relatively small roadside fires from developing into major ground cover fires. In some agencies, structural engines assigned to interface areas commonly carry ground cover hose packs and hand tools.

Figure 5.2 A progressive hose lay consists of extending hose from the apparatus to the fire's edge and extinguishing fire as the hose is extended.

One of the most effective and efficient uses of structural fire engines is for the protection of structures exposed to an approaching ground cover fire. When structural triage (sorting) identifies a structure or group of structures as being defensible (tenable), an engine company can be assigned to protect the structures from the approaching ground cover fire. The following factors limit the usefulness of some structural fire engines for ground cover fire attacks:

- Lack of pump-and-roll capabilities (ability to move while pumping water)
- Lack of four-wheel or all-wheel drive capabilities
- Inadequate approach and departure angles (front and rear overhangs too long)
- Limited water tank capacity
- Limited ability to travel on rough terrain
- Excessive weight for off-road use

All fire departments should have fire apparatus designed to fight the types of fires they confront most frequently. Few, if any, jurisdictions are without ground cover fire potential. Even those fire departments with only developed land within their boundaries may still fight ground cover fires as a mutual aid resource. To accommodate these functions, dedicated ground cover fire fighting apparatus may be needed. Ground cover fire fighting often requires a rugged, highly maneuverable vehicle that can go where large structural-type apparatus cannot.

Because ground cover fires often move rapidly, ground cover fire apparatus may also have the ability to pump water while the vehicle is moving (commonly called *mobile attack* or *pump* and *roll*). The exact design and size of ground cover fire apparatus vary widely from region to region. These variations are necessary because of the differences in terrain and ground cover fuels found in different regions. However, all ground cover fire apparatus should meet the requirements in NFPA® 1906, *Standard for Wildland Fire Apparatus*.

Resource Typing-Incident Command System (NIMS-ICS)

The Incident Command System (ICS) is the basis for safe and efficient incident scene management. As a result of Presidential Directive 5, the National Incident Management System (NIMS) is required to be used. ICS is a component of NIMS.

The first-arriving emergency services personnel establish the NIMS-ICS, make decisions, and take actions that will influence the rest of the operation **(Figure 5.3)**. The initial decisions must be based on the organization's incident scene management procedures.

Figure 5.3 The first-arriving emergency services personnel establish the NIMS-ICS.

Resource typing is a requirement under NIMS-ICS. References for resource typing can be found at www.fema.gov/resource-management. While there is more agreement than disagreement between the various systems, the differences are significant in some cases. Firefighters should know the classification system that is used in their region/province/state.

Resources are categorized by type definition. Measurable definitions identifying the capabilities and performance levels of resources are the basis for each category. Emergency management and response personnel may apply these definitions to inventory their resources.

> **NOTE:** Much of the following information from Resource Typing and Wildland Engine Types is taken or adapted from the following resource: NIMS Resource Management and Mutual Aid. https://www.fema.gov/resource-management-mutual-aid

Resource Typing

Resources may be classified by kind. Resource kinds are broad classes that characterize like resources. The NIMS resources include the following kinds:

- Teams
- Supplies
- Aircraft
- Equipment
- Vehicles

Resource typing is a continual process designed to be as simple as possible to facilitate frequent use and accuracy in obtaining needed resources. To allow resources to be deployed and used on a national basis, the new product development (NPD) is responsible for facilitating the development of national guidance for the typing of resources and ensuring that these typed resources reflect operational capabilities.

Type specifically defines the level of capability a resource has. Type may vary by power, size, or capacity. Therefore, assigning a Type 1 label to a resource implies that it has a greater level of capability than a Type 2 of the same resource. The National Resource Typing definitions are broken into four distinct types. In some cases, a resource may have less than or more than four types. The type assigned to a resource or a component is based on a minimum level of capability described by the identified metric(s) for that resource.

Resource typing ensures that the Incident Command requests, receives, and deploys the resources it needs. Typing also ensures that emergency management and response personnel have the correct definitions available to request and/or deploy the correct resources to the incident.

A number of different types of systems are in use for classifying resources such as engines, dozers, water tenders, air tankers, and crews. However, the minimum standards that each system uses to define the various resource types vary within each category. These resource classification systems allow Incident Command personnel at ground cover fires and incidents of any kind to know what types of resources to request in order to achieve specific fire suppression or related goals and objectives.

Regardless of which system is used, all resource typing deals with *minimums*. Many engines of all types may respond with more than the minimum requirements in personnel, hose, pump capacity, and tank size. In all systems, the various kinds of resources are classified by number according to the minimum standards established for each type. In some systems, the lower the number assigned to a resource, the higher its capabilities are.

Wildland Engine Types

A Type 3 fire engine can be found in a mountainous or rural community. This four-wheel drive apparatus is designed for rapid deployment, pickup, and relocation during ground cover fires (**Figure 5.4**).

Figure 5.4 A Type 3 fire engine.

Type 3 and Type 4 engines are most commonly referred to as *brush* or *wildland* engines because of their lightweight chassis, maneuverability, off-road capability, and smaller diameter hose.

Type 3
A wildland engine with a minimum pump capacity of 150 gpm (600 L/min):
- 500 gallon (2 000 L/min) pump
- 500 feet (150 m) of 1½-inch (38 mm) hose
- 500 feet (150 m) of 1-inch hose (25 mm)
- Gross vehicle weight rating (GVWR) generally greater than 26,000 pounds (11 790 kg)
- A minimum crew of three

Type 4
A wildland engine with a minimum pump capacity of 50 gpm (200 L/min)
- 750 gallon (3 000 L) or more tank
- 300 feet (90 m) of 1½-inch (38 mm) hose
- 300 feet (90 m) of 1-inch (25 mm) hose

- Gross Vehicle Weight Rating (GVWR) generally greater than 26,000 pounds (11 790 kg)
- Requires a minimum crew of two

Type-5, Type-6, and Type-7 engines have at least some fire fighting capability and in many areas they are referred to as **patrols**.

Type 5

A standard duty vehicle chassis:

- Small 10 gpm (40 L/min) pump
- 50-to 200-gallon tank (200 L to 800 L)
- 200 feet (60 m) of 1-inch (25 mm) hose
- Multipurpose unit used for patrol, mop up or initial attack, with a crew of two

Type 6

Initial attack wildland engine with a minimum pump capacity of 30 gpm (120 L/min):

- 150-to 400-gallon tank (600 L to 1 600 L)
- 300 feet (90 m) of 1½-inch (38 mm) hose
- 300 feet (90 m) of 1-inch (25 mm) hose
- Gross Vehicle Weight Rating (GVWR) generally less than 26,000 pounds (11 790 kg)
- Requires a minimum crew of two

Type 7

A standard duty vehicle chassis:

- Small 10 gpm (40 L/min) pump
- 50- to 200-gallon (200 L to 800 L) tank
- 200 feet (60 m) of 1-inch (25 mm) hose
- Multipurpose unit used for patrol, mop up, or initial attack with a crew of two

In addition to these minimum requirements, other basic features are also specified for each of these types of engines.

The same relationship between the number assigned to a resource and its capability holds true for all kinds of resources in all systems. Type 1 dozers are larger and more capable than Type 2 dozers, which are larger and more capable than Type 3 dozers, etc. The minimum staffing requirements (crew of four, etc.) refer to the number of personnel who must be maintained on a 24-hour basis. There are no minimum requirements for designating apparatus equipped with Class A foam capabilities. The Incident Commander (IC) has to specifically request apparatus with foam capabilities if they are needed on the fireline.

Fire Control Tactics

After size-up and the plan has been developed, the Incident Commander must decide where and how to attack the fire. Fires are generally attacked where they are most likely to escape or where they may have the greatest threat to life or property. In the case of an interface situation (explained later in this chapter), the IC needs to evaluate whether to try and contain the fire first to protect a structure or group of structures.

The attack may require resources to attack the heel, flanks, and head of the fire or a combination of all three. This decision is usually made by evaluating the fire's intensity level and rate of spread. A good way to do this is by observing and evaluating flame length. Flame length is a measurement from the mid-base of the advancing flame to the tip. Based on the fire's intensity, rate of spread, and the resources available and their ability to access the fire, the IC can give clear direction as to where and how to attack the fire. Once directions have been given, the crews must select a safe anchor point and begin the attack.

Flank Attack

Flanking fire suppression, also referred to as a flank attack or *flanking the fire*, is used for moderately intense fires moving at a moderate rate of spread. The attack is started at a secure anchor point on one or both flanks of a fire and works toward the head **(Figure 5.5)**. The flanks may be attacked simultaneously or successively, depending on fire conditions and resources available. The attack may be either direct or indirect, and the distance of the control line from the fire edge usually depends on the line. The fire's edge is burned out as soon as possible during fireline construction. Whenever possible, firefighters work in the black for safety.

Pincer Attack

A **pincer attack** is a simultaneous attack on two sides of the fire from a secure anchor point, such as the heel or point of origin, working toward perimeter control. A pincer attack is similar to a flank attack and requires two or more crews working in a highly coordinated effort **(Figure 5.6)**. This coordination demands effective communication between the participating units, especially if there is a mix of air attack, hand crews, and mobile units. Again, whenever possible, crews work in the black for safety.

Figure 5.5 A flank attack.

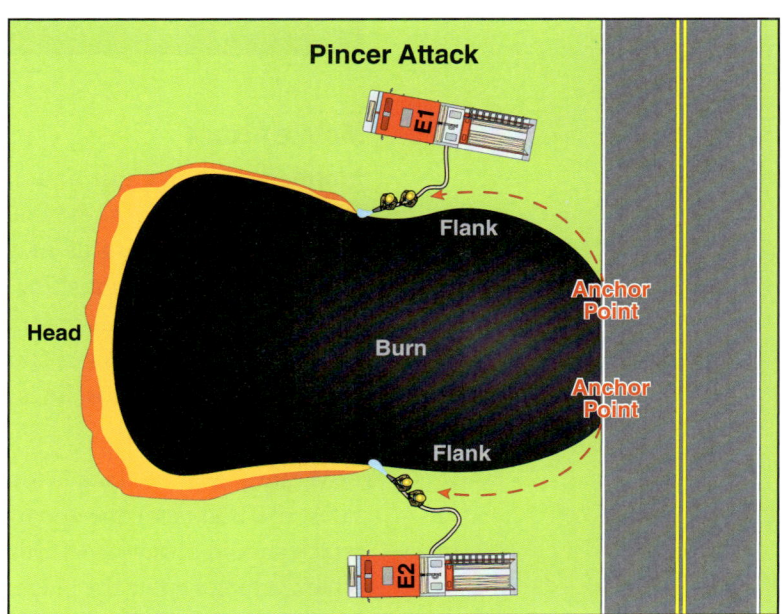

Figure 5.6 A Pincer attack.

Ground Cover Engine Operations • Chapter 5   **109**

Frontal Attack

A frontal attack confronts the head of a fire or the head of fingers extending from the main fire. Because this attack is made without an anchor point, frontal attacks can be very dangerous. While attacking the fingers of a fire is always a high priority in order to slow the rate of spread, it is usually done from within the black for safety. The frontal attack starts at or near the head of a fire and then proceeds to the flanks **(Figure 5.7)**. A frontal attack does limit the spread of a fire; however, such factors as fire intensity, fuel type, wind, or topography often make a frontal attack too dangerous to attempt except by aircraft. If a fire is too intense for a safe and effective frontal attack with ground resources, a flank attack may be the best alternative.

Figure 5.7 A frontal attack may be needed to protect structures.

Mobile Attack

Mobile attack is a fast and efficient method of extinguishing ground cover fires when conditions allow its use. Primarily used on grass fires, it involves employing fire apparatus in a fast-moving direct attack with water or foam as the primary extinguishing agent **(Figure 5.8)**. The major requirement for this method of attack is that the fire be in terrain that the apparatus can safely negotiate. Mobile attack is started from a secure anchor point. Ground sweep nozzles are used, or hoselines are kept as short as possible to allow the engine maximum mobility.

While a mobile attack is sometimes made with the vehicles in the green, the safest method is for the apparatus to work inside the black. In either case, the attack should always start from a secure anchor point. Typically, one or more attack vehicles enter the black at the heel and attack the fire along the burning edge from within **(Figure 5.9)**. This allows the attack to be made directly on the head or flanks of a fire from the relative safety of the black. Because the majority of fuels within the black have already burned, there is less danger of crews being overrun by fire. Crews working in the black are generally exposed

Figure 5.8 A mobile attack.

Figure 5.9 Mobile attacks should be made from the black for safety.

to less heat and smoke. Apparatus operators can see obstacles, such as logs, stumps, and ditches, that might otherwise be hidden if they were in high grass or brush.

Although the prevailing wind tends to blow the smoke away from firefighters, smoldering material within the burn can still produce enough smoke to obscure vision. Therefore, operators and firefighters should:

- Turn on vehicle lights and roll up windows.
- Maintain good communication to reduce the chances of firefighters being run over.
- Rotate the crews more frequently because the smoke can also cause respiratory difficulties.
- Be aware when operating within the black. Do not stop vehicles on smoldering or burning materials that can severely damage tires or undercarriages.
- Use combination nozzles (capable of producing both a fog stream and a straight stream), which are the nozzles of choice in many departments for mobile attack. However, they lack the reach and penetration of smooth bore nozzles. What combination nozzles lack in reach and penetration, they make up for with the spray pattern capability that increases firefighter safety.

The attack lines and their nozzles must have a high enough output to absorb the amount of heat produced by the fire. While "hard lines" (booster lines) are very durable and highly maneuverable, their relatively high friction loss means they may not be able to flow the required quantity of water for medium- to high-intensity fires. In most cases, personnel should use larger lines to ensure safety and productivity.

WARNING
Never ride on the outside of a moving fire apparatus, especially during a mobile attack. Major injuries may result.

Ground Cover Engine Operations • Chapter 5

Tandem Attack

A *tandem attack* is a direct attack by engines, mechanized equipment, hand crews, or aircraft working together along a part of the fire perimeter to achieve greater effectiveness. For example, a tandem attack may be made by aircraft and dozers, dozers and engines, engines and hand crews, or any combination of two resources working in a coordinated effort. The first units do a quick knockdown of the fire; the other units follow closely and do a more thorough job of extinguishment and mop-up.

Two Engines

When two engines attack in tandem, the following occurs:

- The first engine can move along a fireline at a relatively fast pace knocking down a majority of the intensity, knowing that the second engine will be securing the line behind it.

- The second engine moves more slowly, making sure that the fireline and any hot spots along it are completely extinguished (**Figure 5.10**). The first engine completes its assigned portion of the fireline before the second engine.

- The first engine can then either reverse its direction and do a more thorough extinguishment as it works back toward the other engine or begin to extinguish hot spots burning in the black (**Figure 5.11**).

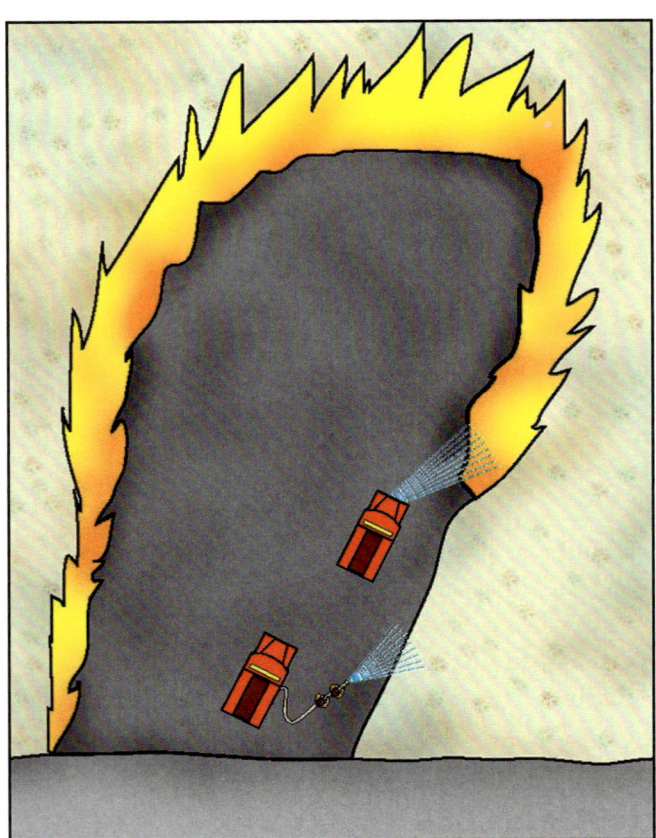

Figure 5.10 Engines working in tandem.

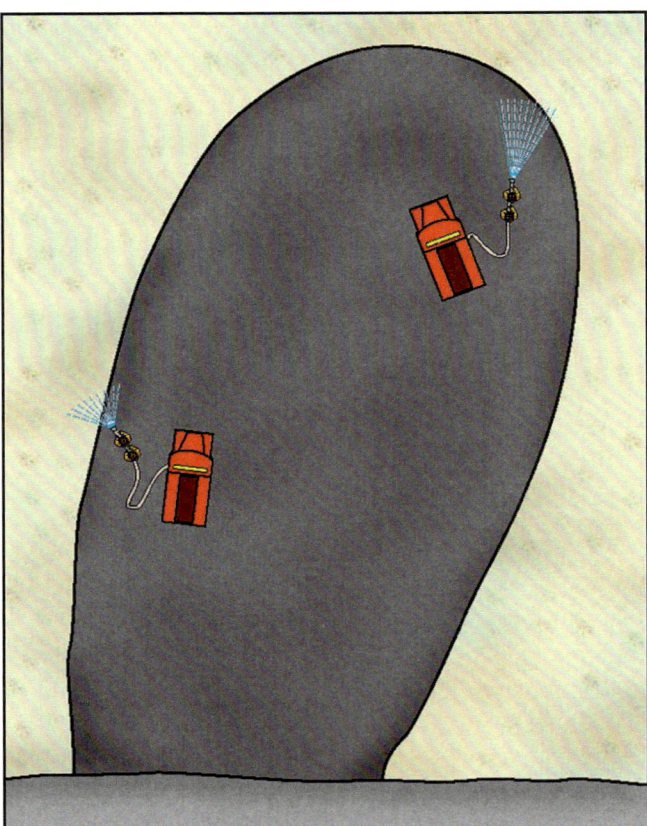

Figure 5.11 The first engine works back toward the other.

Hand Crews

Two hand crews can work effectively in tandem in the same way as two engines. Although crews may progress slower than engines, they are no less effective. A **hand crew** is a number of individuals who have been organized and trained and are supervised principally for operational assignments on an incident (**Figure 5.12**). Because the work of the lead crew may be more physically demanding, crews should alternate taking the lead. A single hand crew is also effective in securing a fireline created by a mobile attack engine. The engine knocks down the fire, while the hand crew suppresses any remaining hotspots or creates a scratch line and picks up any slopover.

Figure 5.12 A hand crew working digging a line. *Courtesy of Kari Greer, USFS. InciWeb-Incident Information System. NWCG.*

Hotspotting

Hotspotting is checking the spread of fire at points of more rapid spread or special threat. It is usually the initial step in prompt control, with emphasis on first priorities. Hotspotting is one of the more dangerous fire fighting tactics. Primarily, it is very dangerous because it is not initiated from an anchor point, and it occurs before control lines are constructed. Hotspotting involves making a rapid attack on hot-burning points on a fire's edge — fingers that are developing rapidly or active parts of a fire that are threatening life or high value property. Only the **hotshot crew** should be assigned because they are an intensely trained fire crew used primarily in hand line construction (Type-1) (**Figure 5.13**). This is a potentially dangerous tactic; therefore, only experienced firefighters should be assigned to hotspotting. Use any available resources for hotspotting. Whenever possible, personnel should use hotspotting from within the black or other natural barriers.

Figure 5.13 A hotshot crew.

Indirect Attack

Indirect attack sacrifices a certain amount of vegetation or other exposed values. Indirect attack is used when:

- The intensity of a fire makes direct attack unsafe.
- The fire develops long fingers.
- There is not enough time to construct a control line at a fire's edge.

During an indirect attack:

- Withdraw fire suppression forces to roads, trails, and other natural fuel breaks or to a constructed control line. This tactic is also of great value in the early stages of initial attack, and resources are scarce or water supplies are limited for direct attack.
- Relocate engine crews to man-made or natural barriers, burn out along those barriers, and hold the barrier with limited resources if they progress at a slow and steady pace.
- Identify safety zones or new ones constructed.
- Notice that the fuel between these barriers and the fire is burned out or **backfired**.

Engines generally support these indirect operations by providing holding support through hose lays and patrolling for spot fires or slopover. While indirect attack does not force firefighters to work as close to a fire front as does direct attack, an indirect attack is not without risk. Firefighters can be in jeopardy during indirect attacks for the following reasons:

- While the control line is being constructed, the fire is growing in size and intensity.
- The line is constructed some distance from the fire's edge; therefore, firefighters are often unable to directly observe the behavior of the fire. This increases the need for LCES.
- It is always risky to have firefighters work in an area with unburned fuel between them and the main fire.
- Whenever firefighters are engaged in constructing or holding an indirect control line, they must be especially mindful of the 18 "Watchout!" Situations.

In an indirect attack, a control line is constructed or located some distance from the edge of the main fire. The distance from the control line to the fire's edge depends on the following:

- Fire's intensity and rate of spread
- Type and volume of fuel
- Topography
- Wind
- Availability of natural barriers.

This attack strategy may be useful if a fire is burning too intensely or spreading too rapidly for firefighters to work safely and effectively at the fire's edge.

A completed control line serves as a barrier to prevent further spread of a fire should it flare up after bringing it under control. Guidelines for a control line:

- The location and construction of the control line determines the effectiveness of an indirect attack.
- The location selected must be far enough ahead of the fire so that the line can be constructed and burned out or backfired before the main fire reaches it.
- The line must be wide enough to prevent oncoming flames from crossing over it to the uninvolved fuel on the other side.
- The crews building the line should always start from a secure anchor point to prevent the fire from outflanking and potentially surrounding them.
- In steep terrain, control lines are usually constructed along ridgelines.
- Ideally, lines should be constructed in light fuels and should take advantage of natural barriers (**Figure 5.14**). An indirect control line should be no longer than crews can control.
- Regardless of the distance between a fire front and the control line, all unburned fuel between these two lines *must* be removed, usually by backfiring or burning out.
- Removing the fuel creates an effective and safe control line; unburned fuel can allow a fire to threaten the control line.

Advantages of an Indirect Attack

One advantage of an indirect attack is that it may make better use of natural firebreaks such as lakes, cliffs, or roads. Also, constructing a control line is generally not as physically taxing on the crews because they are working away from the heat and smoke at the fire's edge.

Figure 5.14 A control line should take advantage of natural barriers.

Disadvantages of an Indirect Attack

Among the disadvantages of indirect attack is that the fuel left inside a control line can allow a fire to increase and reach such intensity and rate of spread that it could jump the control line. The unburned fuel between the main fire and the control line can also allow the fire to develop a large convection column, increasing the risk of spot fires. This prospect is increased if the winds suddenly pick up after the attack is underway. Changes in wind direction, fuel types, and topography could also change the direction of fire spread, rendering the control line ineffective. These dangers reinforce the importance of continually sizing up and adapting to changing conditions. More fire entrapments and fatalities occur on indirect attack than on direct attack. Therefore, indirect attack is generally considered more dangerous.

Hose Lays

An important element in controlling ground cover fires is a **hose lay** that is used either to make a direct attack or to support an indirect attack. A hose lay consists of lengths of hose and accessories laid from a pump and extends to the

Figure 5.15 This hose lay is being used to make a direct attack.

point of water delivery **(Figure 5.15)**. It can be used either as the primary means of suppression or as one of several means in the overall suppression effort. The variables — fuel, weather, topography, and the specifics of the assignment — determine the techniques used.

Booster Line

A **booster line** or hard line may be used for attacking a small, very low-intensity fire or mop-up **(Figure 5.16)**. However, because friction loss in a booster line is high, it delivers a relatively limited volume of water. These hose lays should be no more than a few hundred feet (meters) long and should not be used on large or intensely burning fires where fire-stream effectiveness and personnel safety require the use of larger diameter hose. On actively burning ground cover fires

Figure 5.16 A booster line delivers a limited volume of water.

Figure 5.17 On burning ground cover fires, attack hose must be used to minimize friction loss and deliver maximum volume.

or where long hose lays are needed, at least 1½-inch (38 mm) attack hose must be used to minimize friction loss and deliver maximum volume (**Figure 5.17**). As previously mentioned, the hoselines must deliver enough water to absorb the heat generated by the fire or to protect the crew in case of a flare-up. To provide adequate personnel protection, attack lines may incorporate tee-valves to facilitate lateral extension with smaller hose such as ¾-inch (19 mm) or 1-inch (25 mm) forestry hose. These smaller hoselines are adequate for mop-up within the black, but create too much friction loss to be safe and effective for fire attack.

Progressive Hose Lay

One very effective technique for extended hose lay operations, a *progressive hose lay*, is used primarily for a quick attack on a fire from a secure anchor point, usually a road. It allows a fast, aggressive attack while maintaining a continuous water supply without having to take an engine off the road. A progressive hose lay consists of a hose lay in which double shutoff wye (Y) valves are inserted in the main line at intervals and lateral lines are run from the wyes to the fire edge, thus permitting continuous application of water during extension of the lay (NWCG). Most hose packs contain two, 100-foot (30 m) lengths of 1½-inch (38 mm) hose with lightweight couplings and one hoseline tee with a valved male branch (1½ by 1 inch [38 mm by 25 mm]). Some departments include a 100-foot (30 m) length of 1-inch (25 mm) forestry hose in their hose packs; others keep the 1-inch (25 mm) hose on their engines until needed for mop-up.

To be effective, progressive hose lays require coordination and teamwork. To achieve this, use the following guidelines:

- Start at the heel, attacking the most active flank.
- Extend the hose. When the hose is extended as far as it will reach, use the stream to extinguish the fire.

- Shut down the nozzle, clamp the hose, and remove the nozzle (**Figure 5.18**).
- Roll out an additional length of hose, and attach it to the end of the hose lay.
- Install at any point a hoseline tee with a valved male branch where two lengths of hose are connected (every 200 feet [60 m]) is recommended) if significant mop-up is anticipated or there is a fire threat to the hose lay.
- Screw the nozzle onto the new length of hose. The hose clamp is released, charging the additional hose.
- Repeat this process until either the hose is fully extended or until the fire is completely encircled. Additional personnel may be needed to help pull hose, to extinguish hot spots, and to help with mop-up.

Figure 5.18 Hose is clamped as another section is pulled from a pack.

Ground Cover/Urban Interface Operations

Almost all fire departments face the problem of protecting the ground cover/urban interface. This is defined as where man-made improvements are in contact with ground cover fuels. These improvements are not only structures but power lines, pumping or electrical stations, oil and natural gas infrastructure, recreation sites, research stations, and the like. Scenic beauty and changing lifestyles have motivated people to purchase and develop homes and businesses in once pristine areas. People and their homes and workplaces increase the ground cover fire problem in a number of ways. Fire protection planners must be concerned with the increased life hazard that the larger populations in these areas create. The increased human activity multiplies both the ways in which fires can start and the total number of fires in these areas. Structures in the ground cover greatly raise the dollar value of property to be protected.

Because of increased human habitation in the ground cover/urban interface and the resulting growth in the values at risk and the number of fires, departments that never planned to function at ground cover fires have been asked to respond to mutual aid calls in the interface. Therefore, it is critically important for all firefighters — from those in single station volunteer companies to those in large metropolitan departments with dozens of stations — to be able to perform safely and effectively in these fires.

Command, Control, and Accountability

Probably the greatest challenge for operating in this environment is establishing a control and accountability system that works when the Commander almost certainly will have limited, if any, visual confirmation of where his or her crews and companies are working. Having a vast number of differing resources, structural engine companies, ground cover engine crews, hand crews, or heavy equipment working in two totally different environments at the same time can be daunting. This can be overwhelming when combined with the pressures of the potential of the threat to the life safety of firefighters and citizens of the area as well as high-volume property loss. Several viable options have been proven time and again in these situations that any first-arriving Incident Commander can employ for success.

The first of these has been described in Chapter 3 and that is performing a timely and complete size-up. This cannot be underscored enough. In most interface situations, our first inclination as firefighters is to rush in and take action because of the threat to life and property. While action is certainly required in a timely fashion, deploying limited resources without organization is one sure way of raising the risk to personnel to an unacceptable level. Some things for consideration in your size-up are:

- Size of the fire (acres involved)
- Type of fuel involved (light, medium, or heavy)
- Rate and direction of spread (toward or away from structures)
- Fire behavior (surface, running, crowning, etc.)
- Best access/egress: is there more than one way in/out (road name, property name, etc.]
- Life safety (evacuation needed, etc.)
- Exposures threatened
- Hazards (power lines down, bridges that might not support engines, etc.)
- Additional resources needed
- Water supply (hydrants, ponds, and private supplies)

The next course of action is to establish a staging area for ALL incoming resources **(Figure 5.19)**. A staging area is a location set up at an incident where resources can be placed while awaiting a tactical assignment on a three (3) minute available basis (NWCG). This ensures timely and appropriate assignment of the right resource to the right location, and no freelancing can take place. The assignment of a staging area manager to track incoming, assigned, and available resources can be invaluable as things progress. The Staging Area Manager (STAM), the ICS position responsible for supervising a staging area, reports to a Branch Director or Operations Section Chief.

Figure 5.19 A staging area for all incoming resources. *Courtesy of FEMA.*

Before committing any resources, you must organize your resources:

- Locate safety zones.
- Post lookouts where they can see the main fire and the structures that are threatened.
- Determine a trigger point that the lookout will utilize as the key to informing all working companies and resources to evacuate to those predetermined safety zones.

As initial actions are clearly necessary, there are two (2) ways to organize your initial attack resources.

1. Divide your incident into specific geographic locations to which you can assign resources (commonly referred to as Divisions) under the supervision of a *Division Supervisor* or organize them into functional Groups under the supervision of a *Group Supervisor* (Examples of functional groups in this scenario might be a structure protection group, firing group, or fire suppression group).
2. Formulate the resources in staging into either Strike Teams under the direction of a Strike Team Leader or Task Forces under the direction of a Task Force Leader, assigned to the divisions or groups.

Strike Team

Strike teams are generally five like resources with common communications and a leader (such as five Type 4 brush trucks [the exception to this is dozer and crew strike teams only have two]).

Task Force

Task forces are generally five different resources with common communications and a leader (such as two Type 4 brush trucks, two Type 1 structural engines, and one Type 2 hand crew). This configuration gives you good accountability and the most "bang for your buck" in most interface situations.

Ingress and Egress

Initial attack Incident Commanders and incoming resources must take note of and share information on the best routes to access or egress from affected or threatened areas of the incident. Much of this information may be available to the local jurisdiction through preplans and familiarity from day-to-day operations or previous fires, but not so for outside resources. Mutual aid resources must be made aware of the safest and timeliest means of access and egress to complete the mission. Before committing personnel, the following information must be considered:

- Is there only one way in/out?
- Are the roads too narrow for larger structural protection apparatus? If they can access the area, is there ample room for them to turn around in cul-de-sacs, yards, and dead-end streets; or will they have to back in?
- Are the roads mid-slope with heavy fuels below and above? Are they one way? If they allow for two-way traffic, can evacuating residents and incoming fire apparatus pass each other safely?

- Are there heavy fuels on both sides of the road that once were affected by fire and would limit escape from the area due to radiant heat?
- Are there power lines along or crossing over the access/egress routes that if impacted may preclude access or escape from the area?
- Does the access route dead end? If so, is there ample room to turn apparatus around?

Residents and the Public

Some residents are ready to flee at the first sight of smoke. Others will want to stay at their homes. You will need to provide advice and direction. The two most common strategies for dealing with residents are **shelter in place** and **evacuation**. Residents that remain can be helpful; they may know the locations of other structures, water sources, access routes, hazards, etc. They can help prepare their home before the fire hits.

Advise homeowners who remain on the following basic safety considerations:

- Be alert to equipment.
- Do not go out into unburned fuel.
- Know the escape routes and safe zones.
- Remain in the structure if trapped by the fire. Exit when it is safe to do so.

Evacuation

Evacuation is usually the responsibility of law enforcement agencies. Each state and legal jurisdiction may have different laws. For example, Wyoming State Statute 35-9-116 states in part, "In the event of a hazard of immediate life-threatening severity, the state fire marshal or the chief of a fire department or district may order evacuation of a building or area and may implement emergency measures to protect life and property and to remove the hazard." In other jurisdictions, only the highest elected official or county emergency management official may authorize an evacuation order. Every firefighter and officer should know what the laws pertaining to this subject are in their home jurisdiction.

Figure 5.20 Law enforcement has the authority to require people to evacuate their homes.

Evacuation may be required to clear the area for fire fighting operations and to minimize risk to citizens. We can ask people to evacuate, but only law enforcement officers have the authority to make them leave **(Figure 5.20)**. Advise evacuees to take a minimum of belongings with them. Suggest they close-up, but not lock their residences. Direct them to the appropriate evacuation route or shelter location and to watch for incoming equipment.

Routing Traffic and Establishing Access

Request assistance from local law enforcement for traffic control. If law enforcement is not on scene, delegate traffic control. Use the following procedures:

- Use flares, emergency lights, and other visible safety warning devices at all times.

- Coordinate traffic control with law enforcement when they arrive on scene. You may encounter narrow access roads already filled with, and even blocked by, local traffic.
- Develop a traffic plan and communicate the information to all units and dispatch.
- Identify routes into and out of the area with signs or flagging.
- Clear existing traffic to make way for fire equipment. Alternatively, direct civilian traffic to the roadside until fire equipment has passed and tell them when they can move out.
- Leave a clear path for other incoming resources.
- Note weight limits or bottlenecks that may limit some equipment.

Structure Triage

Structure triage is sorting and prioritizing structures requiring protection from ground cover fire. Triage may be required of anyone at any time on the incident, from the Incident Commander doing reconnaissance to the engine crew moving into position.

Firefighter safety must always be the first consideration of structural triage. The goal of triage is to do the best with what you have and to not waste limited resources or time. It requires that you quickly categorize threatened structures.

Structure Triage Categories:
- Needs little or no attention for now
- Needs protection, but savable
- Cannot be saved

There are no fixed answers based just on the structure itself; no one can look at a house and surrounding fuels alone and choose the category that will always apply. The decision process must be consistent and logical in order to reach a decision based on all the relevant factors.

Greatest Potential Threat

Look at the greatest potential threat, which is based on the assumption that the fire behavior will be the worst possible under the prevailing conditions. While you may not base your actions on such a possible threat, at least have an alternative plan should the worst develop. Consider the following:

- Fuels (in your estimation of their driest condition), firebrands, worst weather that might occur, and terrain
- Greatest vulnerability of the structure (construction features, materials, proximity to fuels, and position of the structure relative to the topography)

Probable Threat

Look at the probable threat, which is based on the fire behavior that is most likely to occur under the conditions. It is this situation that should guide your decision on the action to take. Consider the following:

- Actual fire intensity and firebrand problem that you expect
- Aspects of the structure that remain vulnerable under the expected fire behavior even with some clearing and protective action being taken

- Arrival of the fire and how long the threat to the interface lasts

The probable threat will determine preparation and commitment time. When other resources arrive will determine their usefulness. Consider the following:

- Rate of spread and intensity.
- Orientation dynamics of the fire as it moves into the structures.
- Arrival times of other resources.
- What can be done with the resources that are available?

This has to be your best judgment of what you can accomplish in the face of the expected threat. You must reach your decision on where to put your effort. One approach is to imagine the effect of putting all required resources on the most threatened structure. Based on that outcome, look at the effect of shifting resources to other, less threatened structures. In the final analysis, you want to save the most structures. If the most threatened structure cannot be saved, forget it. Then access the next most severely threatened structure. If it cannot be saved either, then move to the next, most threatened, etc. If a threatened structure can be saved, you must still decide if that is the best thing to do. Even though you save one, the effort might cause you to lose others that could have been saved. Instead, ask what will happen if resources are applied to less threatened structures. If you can then save only a different structure, but no more than one, go for the tough ones. If, on the other hand, you can then save two or more structures, drop the more threatened ones. Continue the "what if" process until you feel you are at a point where you can save the most structures with the help you have.

Factors Affecting Triage

The following factors affect triage decisions:

1. Structure
2. Fuel
3. Fire Behavior
4. Resources
5. Firefighter Safety

1. Structure. Are the structure and exposure susceptible? Construction features and conditions to evaluate are:

- **Roof**
  — Combustible: wood shakes, tar paper, etc.
  — Not combustible: tile, metal, or fiberglass, etc.
  — Pitch: debris on roof or in gutter
- **Siding**
  — Combustible: wood
  — Noncombustible: metal, brick, etc.
- **Heat traps**
  — Open gable
  — Vents without screens or non-fire-resistant screens
  — Overhanging decks

- **Windows**
  - Large area windows, single/double pane
  - Non-fire-resistant screens
  - Position on the house relative to the direction of oncoming fire threat
- **Size of building**
  - Can single resource protect using standard practices?
- **Shape of building**
  - Large flat surfaces vs. numerous corners/heat traps
- **Position on slope**
  - Mid-slope, poor access
  - Top of slope, vulnerable to radiant or convective heat, apparatus will be exposed on ridgelines
- **Surrounding fuels (defensible space)**
- **Fire behavior**
  - Available resources
  - Firefighter safety

2. Fuel

- **Size and arrangement**
  - Light/flashy
  - Heavy dead and down
  - Continuous vs. patchy
  - Surface, ladder, canopy
- **Age**
  - Early season
  - Late season cured
  - Disease/frost kill
- **Proximity**
  - Within 30 feet (9 m) of the structure
  - Overhanging or in close proximity to access/egress routes
- **Types**
  - Resistant or flammable
  - Landscape/ornamental
  - Grass, brush, timber, exotic (palmetto, etc.)
  - Wood piles
- **Landscaping**
  - Railroad ties, cedar (wood) fences
- **Defensible space, access**
  - Open/green yard
  - Wide/narrow drive, space to maneuver/position apparatus

- **Yard accumulation**
  - Free from clutter; can personnel easily maneuver hoselines
- **Explosive: LPG tanks, diesel or gas storage tanks**
- **Other hazardous materials, vehicles, etc.**

  3. **Fire Behavior.** Fire behavior refers to how the fuels will burn.

- **Rate of spread and direction**
  - Running/head fire
  - Backing/flanking
  - Fuels produce numerous firebrands for spot fires
- **Topographic influence**
  - Daytime, slope/valley winds
  - Narrow/box canyons
  - Man-made wind tunnels (tightly spaced structures on narrow streets)
- **Weather influence**
  - Relative humidity below 25%
  - Temperature above 85 degrees F (49 degrees C)
  - Unstable atmosphere
- **Flame length**

  The distance between the flame tip and the midpoint of the flame depth at the base of the flame (generally the ground surface), which is an indicator of fire intensity (NWCG).
  - Less than 4 feet (1.2 m) susceptible to direct attack operations.
  - 4 to 8 feet (1.2 m to 2.4 m) can be directly attacked with mechanical assistance (engines/heavy equipment).
  - Greater than 8 feet (2.4 m) indirect or defensive attack strategies may be necessary.

  4. **Resources.** Resources refers to what is available and when.

- **On-site resources**
  - Water onsite: wells, swimming pools, stock ponds
  - Hand tools, ladders, and equipment that may be found in the shed/barn on the property
- **Kind and type of equipment available**
  - Structural vs. ground cover engines
  - Heavy equipment, skid steers, dozers, motor graders, tractors/plow/discs
- **Number (sufficient initial attack, properly staffed)**
- **Where they are (committed, staging, en route)**
- **Time of availability and response time**
- **Capabilities and limitations**
  - On-road/off-road, sufficient tank capacity for sustained structure defense

- Mobility
- Water/foam/retardant

5. Firefighter Safety

- **Ingress/egress routes**
  - Adjacent fuels
  - One way/two way
  - Canopy
  - Slope and steepness of road
  - Loops
- **Power lines**
- **Smoke/visibility**
- **Hazardous materials**
- **LPG and fuel storage tanks**
- **Many others (remote wooden bridges, fire crossing road, etc.)**
- **Structural Situations that Shout "Watchout!"**
  - Wood construction (shake/shingle roofs, elevated decks, etc.)
  - Poor access (narrow, one-way roads)
  - Inadequate water supplies
  - Flammable vegetation within 30 feet (9 m) of structures
  - Extreme fire behavior
  - Strong winds (in excess of 25 mph [40 km/h])
  - Emotional residents (angry, panicky, or hysterical)
  - Structures situated in natural chimneys or box canyons, on slopes in excess of 30 percent, in flashy fuels, etc.
  - Weak or narrow bridges
  - Open eaves and soffit vents
- **The Dos and Don'ts in Protecting Structures in the Interface**
  - Wear full protective clothing and equipment.
  - Reserve at least 100 gallons (400 L) of water for engine/crew protection.
  - Have a charged line for engine/crew protection.
  - Back engines into position for quick exit if necessary **(Figure 5.21)**.
  - Use at least 1-½ inch (38 mm) hoselines whenever possible.
  - Post lookouts whenever necessary.
  - Have an escape plan.
  - Do not park in saddles or chimneys.
  - Do not attempt interior structural fire fighting unless trained and equipped to do so; no other structure protection priorities are evident; and the effort has been approved by the Division/ Group Supervisor or Task Force/Strike Team Leader

Figure 5.21 Engines should be backed in from the last known turnaround.

Consider All Factors

Triage is a logical process. It is not an answer or simple formula. Triage requires you to make basic predictions of fire behavior and to estimate the capabilities and availability of resources. You must base your decisions on probabilities — play the odds. Several triage checklists have been developed for use by homeowners and firefighters.

When Structures Cannot Be Saved

No simple rule will tell you when to try or what time to abandon a structure defense effort. The following factors or conditions are worth noting. If any of these apply, then the attempt to save that structure deserves careful consideration before continuing.

- The fire is making significant runs (not just isolated flare-ups) in the standing live fuels; for example, the brush or tree crowns and the structure are within one or two flame lengths of those fuels.
- Spot fires are igniting around the structure or on the roof and beginning to grow faster than you can put them out.
- Your water supply and stream flow will not allow you to continue fire fighting until the threat subsides.
- You cannot safely remain at the structure and your escape route could become unusable (blocked by fire, falling or rolling obstacles, etc.).
- The roof is more than ¼ (6 mm) involved (in windy conditions), and other structures are threatened or involved.
- Interior rooms are involved and windows broken (in windy conditions), and other structures are threatened or involved.

If things change, or if you are losing the battle, rethink your plan, but do not continually question or regret your decisions. Time wasted in indecisions is very costly. This is not a situation that allows lengthy deliberations. The situation does not allow more than a best judgement and a good effort. Make decisive judgements and make them without undue delay. Then go to work.

Structure and Site Preparation

Much can be done to improve the chances of saving the structure if there is any time available before the fire reaches a structure. Site preparation depends upon the time and assistance you have prior to the fire's approach. Initial attack on interface fires offers little time for preparation. Often, all that can be done is to get an engine to the structure and position hose lays. Base site preparation on the fuels, expected fire behavior, and the information you gather conducting structural triage. Use engine crews, hand crews, heavy equipment, and other resources available for structure and site preparation.

Structure Protection: Lessons Learned

The following list presents important information to consider:

- Tactics employed in structure protection are the same for both ground cover and structural fire fighting agencies, regardless of the type of resources utilized.
- Most interface fires occur under high wind conditions, creating rapidly moving fires, with extreme fire behavior, long-range spotting, and multiple fire fronts.
- The scattered location of structures in the interface can limit tactics, such as direct attack or burnouts, commonly used in ground cover fire fighting.
- Spot fires create multiple fire fronts, and firefighters protecting structures are often surrounded by flames, showered by burning embers, and subjected to dense smoke during the battle to save someone's home.
- Escape routes and safety zones are easily compromised in structure defense by remaining at the structure beyond what we would consider safe in ground cover fire operations.
- Mobility is one of the most important tactics employed in structure defense. Engines must be able to quickly move from house to house in the protection effort. Structure engines are larger and less mobile than ground cover engines. Consider actions in the deployment of fire fighting equipment that will allow for rapid response to the changing fire environment while maintaining the ability to escape to a safety zone.
- Wise water use is critical to structural defense. Water may be most effectively used in foam solutions to wet down structural exposures prior to the arrival of the fire front.
- Coordination, organization, and communication may not be adequate during initial operations.
- Resources required may not be available, and those on scene may not be able to control the spreading fire. Resources defending structures must be mobile, resourceful, and self-reliant.
- The ability to communicate among all agencies responding to interface fires is an absolute must. Regular communication among all resources is essential.
- Situational awareness is required due to the numerous factors that can quickly compromise the safety of everyone involved.

The Structure

Look at the structure as fuel. Wood roofs and siding are more vulnerable to ignition than noncombustible types. Virtually any opening into the structure is an entry point for firebrands. Pay particular attention to the likely ignition points:

- Shake roofs
- Cedar lap siding
- Open vents
- Open, broken, and windows without screens
- Open doorways or breezeways
- Open crawl spaces
- On and under decks
- Other flammable materials

On-Site Resources

Look for things that you can use to help prepare the structure and fight the fire. With a little evaluation, lots of things around a home can be put to good use. Such things include:

- Materials for covering openings (plywood, boards, sheet metal, etc.)
- Hammers, saws, nails, wire, etc., for securing coverings
- Ladders (put on the safe side of house) **(Figure 5.22)**
- Rakes, brooms, blowers, etc., for removing leaves, needles, or grass
- Chain saws, trimming saws, axes, shovels

Locate Water Sources

Locate water sources that could be used, even small ones. Such possible sources include:

- Hydrant types
    - Wet barrel
    - Dry barrel
    - Private industrial
- Agricultural hydrants that require activation before use
- Pools
- Cisterns and tanks
- Irrigation systems
- Garden hose outlets (good for filling engine tank)

Figure 5.22 Place a ladder against the eaves of the roof.

Adjacent Resources

Use the following guidelines to maximize your use of other resources:

- Contact fire units adjacent to your area of protection.
- Determine mutual protection boundaries and adjust assignments, if necessary, to even the workload.

- Write down radio call ID's and frequencies.
- Learn the routes to use in moving to assist each other.

Clearance Around Structures

Research indicates the following:

- Large flames and crown fires generally do not ignite homes.
- Intense fires burning farther than 100 feet (30 m) from a structure do not transfer enough radiant heat to ignite the structure.

  More often, the following small ignitions and spotting start structures on fire:
  — Firebrands landing on combustible material of or near the home.
  — Continuous surface fuels allow surface fires to spread to and ignite the structure.

Also, consider the home ignition zone that determines the vulnerability of a home and surrounding area to ground cover fires. The home ignition zone includes the home and extends a distance of 100 to 200 feet (30 m to 60 m) around the outside perimeter of the home.

Removing and Trimming Fuels

State laws vary on who may or may not have the authority to remove fuels around private property. Get permission from the landowner or appropriate local authority. Suppression resources have to communicate, coordinate, and cooperate with the local jurisdictional entities in the interface.

Use the following guidelines for removing fuels around private property:

- Clear combustible material and vegetation from around the structure.
- Make the clearance at least three times the expected flame length in the primary fuels.
- Use discretion and consider the homeowner's efforts and expense in landscaping.
- Use foam to wet down landscape trees and shrubbery adjacent to the structure to protect the home.
- Leave isolated or widely scattered plants and most ornamental shrubs and trees.
- Trim lower branches and eliminate other ladder fuels to effectively isolate the aerial fuel from the fire.
- Pile cleared vegetation where it will not burn or will not cause a problem if it does.
- Do not simply fell trees or lop off branches and leave them lay. This may create a more hazardous fuel bed than you had before.

Fireline Construction

Fireline is a strip of mineral soil cleared of vegetation intended to stop the spread of the fire. Construct a fireline in fuels and terrain where you can control the main fire or your firing operation. Light fuels, grass, scattered shrubs, and forest litter are the best location for fireline construction because of minimizing the amount of work required and decreasing the exposure of firefighters holding the line. Try to use openings in tight forest canopies.

Locate the fireline as close as possible to the structure. If flammable vegetation remains inside the control line, firebrands could still carry fire to and ignite the structure.

Take advantage of existing breaks in the fuel. These can include:

- Roads and driveways
- Lawns and landscaped areas
- Grazed and trampled grass
- Power line rights-of-way
- Trails or paths

Intermediate Fuels

Intermediate fuels are any combustibles located near the structure. They can sometimes convey fire directly to the structure, produce firebrands, or radiant heat that will threaten the structure. Common examples of intermediate fuels are:

- Wood piles such as lumber, firewood, and fencing materials **(Figure 5.23)**
- Wooden fences attached to the structure **(Figure 5.24)**
- Attached decks and combustible awnings
- Combustible yard furniture **(Figure 5.25)**
- Wood swing sets and play houses

Yard Accumulation

As well as the obvious combustibles that can directly threaten the structure, common objects scattered around the yard can create control problems or have a value worth protecting. Yard accumulation can interfere with the placement and movement of hoselines, and it can also greatly complicate and delay firing operations. Common objects can include:

- Inoperative vehicles **(Figure 5.26, p. 132)**
- Boats and small trailers
- Power tools
- Stored material (pipes, poles, etc.)

Figure 5.23 Wood stacked next to a structure may provide more fuel to a fire. *Courtesy of NIFC.*

Figure 5.24 Attached wooden fencing can help a fire spread to the structure.

Figure 5.25 Wooden lawn furniture can provide more fuel to a fire.

Figure 5.26 Items, such as old cars and trailers, left in the yard can provide more fuel to the fire.

Flammable and Explosive Hazards

Many things can burn violently or explode and deserve special attention as soon as possible. For example:

- Elevated gasoline or diesel tanks. Clear fuel around such hazards to a distance adequate to protect them from excessive radiant heat. The required clearance will depend upon fire intensity and your ability to cool or shield them.
- LP gas tanks.
- Vehicle components (batteries, shocks, tanks, mounted tires, drivelines, etc.).
- Pressure vessels and aerosol cans (even if the contents are not flammable).
- Outbuilding for storage of fertilizers, pool chemicals, motor vehicle fluids (diesel fuel, brake fluid, oil, etc.).
- Other hazardous materials.

Structural Exterior and Interior Preparations

The roof is the most readily and frequently ignited part of a structure exposed to ground cover fire. Use the following guidelines for exterior preparations:

- Clear needles and leaves off the roof and out of the rain gutters if it can be done safely.
- Use ladders to access roof areas that cannot be wet down with hose from the ground level.
- Avoid contacting electrical lines with water or when moving a ladder.
- Avoid climbing on roofs if possible. Wet roofs and high winds create the potential for falling.
- Cover openings and potential openings. Any entry of fire or firebrands into the structure greatly increases control problems and the likelihood the structure will be damaged or destroyed.

- Concentrate your efforts to openings on the side of the structure that is exposed to the fire. Leave window screens attached, and close any exterior window coverings.

When preparing the interior of a structure, be sure to do the following:

- Close windows.
- Close nonflammable window coverings such as blinds, shades and drapes.
- Close interior doors to limit fire spread should the interior become involved.
- Turn off fans and swamp coolers that may allow embers into the structure.
- Turn off gas (LPG or natural) at the source (**Figure 5.27**).
- Leave on electricity to run pumps, provide lighting, etc.
- Leave on a porch light and a central interior light to provide visibility in dark, smoky conditions (**Figure 5.28**). Patrolling engines will more easily notice the house, and firefighters entering it will have some light.
- Make sure that essential doors can be opened. Close, but do not lock, all doors.
- Leave a note for the homeowner. Describe in what condition you have left the structure (utilities, pets, etc.).

Figure 5.27 Turn off the gas at the source.

Figure 5.28 Leave on a porch light.

Private Vehicles

Vehicles that will remain on-site can be taken care of to minimize damage to them and to the degree to which they will be in the way. Be sure to do the following:

- Park vehicles in a sheltered location, away from heat and firebrands.
- Make sure that any vehicles will not interfere with the movement of fire equipment.
- Do not park vehicles over flammable vegetation. If flammables are in the area, spray a foam blanket around and underneath the vehicles.
- Park the vehicle headed out, if possible, with the keys in the ignition.
- Close, but do not lock, the doors and windows.

Figure 5.29 Livestock may need to be evacuated.

Pets and Livestock

Most often, animals that are free to move around will manage to avoid being burned. However, if they are fenced or chained they may need to be freed (**Figure 5.29**). Troublesome or frightened pets might need to be placed in the garage, residence, or other enclosure. If a large problem with pets or livestock is encountered, call for assistance from the local animal control agency.

Pretreatment of Structures

Sprinkler Systems

Sprinklers may be used to wet down the structure or the vegetation around a structure. Encourage property owners to turn on their system if there is potential for the fire to reach their property.

Class A Foam

The use of class A foam is a proven technique in protecting structures.
- Can be quickly applied to the structure using engines or portable tanks.
- Is easy to use by batch mixing in tank without foam proportioners.
- Minimizes removal of ornamental landscaping and fireline construction. It can be used to wet down landscape vegetation around the structure.
- Maximizes firefighter safety. Crews move to safety zones until fire front passes and then return to conduct any needed mop-up.

Fire Gel

Fire Gel is produced by commercial vendors under various trade names. Fire Gel is a gel concentrate that when added to water transforms water into a fire-preventing and heat-absorbing gel. It will adhere to any kind of surface, even vertical window panes. Fire Gel is applied by special nozzles and systems.

Structure Wrap

Structure wrap is available from commercial vendors under various trade names. It comes in rolls (approximately 3 feet wide by 300 feet long [1 m by 90 m) and is made from similar material as the fire shelter. It can be reused if care is taken when removing it from the original application.

New materials and chemicals are currently being developed that have proven effectiveness in protecting structures from fires while minimizing the exposure of firefighters. It is important to stay current with rapidly developing technology.

Structure Protection Tactics

Engine crews and apparatus are the primary resources used in structural protection. Positioning the engine is a proven technique in successful interface engine tactics that maximizes efficiency, mobility, and firefighter safety. To make the engine safe and convenient to work from, follow these guidelines:

- Do not block travel routes for other equipment or evacuating vehicles. Park off the road.
- Do not park over flammable vegetation. Scrape or burn away the fuel from your parking area if needed.
- Park on the side of the structure that will minimize exposure of the engine to heat and blowing firebrands.
- Be near enough, but not right next to, the structure to limit the length of hoselines.
- Avoid structure collapse zone (1½ times the height of the structure).
- Avoid parking next to or under such hazards as:
  — Power lines
  — Flammable trees or snags
  — LP gas tanks; pressure valves
  — Buildings that might burn
- Leave the doors, windows, and compartments closed and the keys in the ignition. You do not want to find your vehicle on fire.
- Always maintain a lookout, usually the pump operator, with the engine.

Working Hoselines

In structural protection, 1½-inch (38 mm) hoselines are recommended for use. To protect all sides of the structure, 1½-inch (38 mm) single-jacket forestry hose provides the mobility needed. In fine fuels with low-intensity fires, ¾-inch (20 mm) or 1-inch (25 mm) hoselines can provide a mobile and reliable choice. Even though these smaller hoselines are more maneuverable, they provide much less protection for firefighters in case of a sudden increase in fire intensity. One disadvantage of a hard line is that it cannot be rapidly cut off and abandoned if escape becomes necessary.

When protecting structures:

- Deploy two lines — one around each side of the structure or around a pair of adjacent structures. The lines must be long enough to meet behind the structures.
- Be aware of the pump and water capacity of your engine.
- Keep in mind that Type 6 and Type 7 engines may not have the pump capacity or water supply for extensive hose lengths.
- To rapidly deploy and reload structure protection lines, a 100-foot (30 m) by 1½-inch (38 mm) single-jacket hoseline may be preconnected and secured to the rear of an engine by means of webbing or a strap.
- Ensure personnel safety by not requiring them to climb on top of an engine.
- Make sure that lines have a shut-off valve at the engine. This will allow the lines to be rapidly disconnected should escape become necessary.
- Deploy the lines around and behind the engine, not in front, to prevent the hoses from wrapping around a wheel in a rapid egress situation.

- Use supplementary lines for backup, interior attack, or spot fires on the back side of the structure.
- Charge and check any line positioned for immediate use.

When setting up engine protection lines:

- Partially charge and coil a 50-foot (15 m) section of 1½-inch (38 mm) hose near the vicinity of the engine control panel.
- Set the hose where it can be easily reached and re-charged in the event the fire overruns your engine's position. Make sure that it will not fall off if the engine is moving.
- Protect adjacent structures or retreat to a safety zone if an engine has to leave to refill. Working lines left in place can immediately be put back in service when an engine returns to the scene.

Use the following guidelines for working lines:

- Leave in place until the structure is out of danger.
- Leave so that they are easily noticed and within reach of the outlets on an incoming engine. Mark them with flagging.
- Do not leave the couplings where they might be run over.
- Drape the ends of the lines over a fence, mailbox, etc. **(Figure 5.30)**.
- Lay the working lines before an engine actually takes up a position. For example, hand crews could set up hose lays prior to the arrival of the fire front, saving critical time for the engine units when they arrive.
- Cover the hoselines with dirt for protection from heat.

Figure 5.30 Drape the ends of the lines over a fence, mailbox, etc.

Nozzles

A combination nozzle is generally the most versatile. It provides for conservation of limited water supplies when using the spray tip for wetting down exposures, or the knockdown power and reach of a straight stream. Straight stream tips on working lines or roof lines can provide a better water stream in high wind conditions. Air aspirating foam nozzles or combination foam nozzles, when used with properly mixed foam, provide good pretreatment for structure protection.

Confronting the Fire at the Structure

Strategies and tactics for protecting a structure when the fire front arrives depend upon the type of fuels surrounding the structure and the equipment available. Structures surrounded by fine fuels can effectively be protected by stopping the fire spread with water or firelines. Using water to prevent fire spread in running crown fires with brush or timber fuel types is ineffective in relation to fire intensity and exposes firefighters to undue risk. Water is most effectively used in heavy fuel types with foam application prior to the fire's arrival or for putting out spot fires on the structure after the fire front passes.

Spotting Zone

Most interface fires will put you in the spotting zone (**Figure 5.31**). A spotting zone is behavior of a fire producing sparks or embers that are carried by the wind and which start new fires beyond the zone of direct ignition by the main fire (NWCG). Airborne firebrands are the biggest problem because they can ignite spot fires, and the threat may exist for several hours. Firebrands may ignite new fires as far as a mile or more ahead of the main fire. The main fire may move through later (putting you in a different situation), or it may never get there.

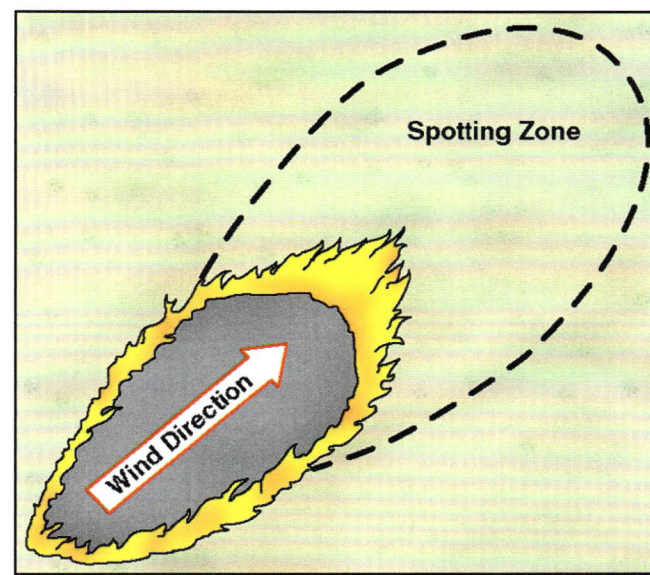

Figure 5.31 Most interface fires will place you in the spotting zone.

- Remain mobile enough to quickly reach any point within your area of responsibility. It may not be necessary to deploy lines except to actually put out a fire.
- Check constantly for new ignitions or receptive fuel beds, including roofs and woodpiles.
- Patrol as necessary.
- Post lookouts with communication.
- Make sure that there are no gaps in surveillance between adjacent areas.
- Attack a spot fire quickly.
- Make sure the fire is completely out or has a control line capable of preventing its spread.
- Remain alert for other spot fires.

Ground Cover Engine Operations • Chapter 5   **137**

Full Containment

Full containment means attacking and stopping fire before it reaches the structure. Light fuels and low intensity fire provide opportunities to prevent the fire from reaching the structure. If you cannot wait for the main fire, or if the fire will be too intense for direct control, you can fire out from a control line. Firing operations and techniques will be explained later.

Partial Containment

If there is not enough time or the fire intensity will not allow you to establish complete containment, you can still attempt to reduce the fire's intensity as it moves towards the structure if you have adequate water supply. If not, save the water for the structure. Use your working lines to knock down the segment of the fire front that is moving directly toward the structure. After the main fire passes, check the structure for possible ignitions, such as on the roof, under eaves, rain gutters, and wood decking.

No Containment Possible

When containment is not possible, the ground cover fire will burn over and past the structure unchecked. In this case, suppression efforts are focused on the structure. Ensure adequate safety zones are accessible, available, and known to all personnel. If you have an adequate water supply, direct all hoselines onto the structure and allow the ground cover fire to burn past. If the fire intensity threatens your safety, retreat to a safety zone and re-enter the area when the fire has passed or coat the structure in Class A foam and leave until the fire front has passed.

Fighting Roof Fires

Ground cover fires frequently ignite combustible roofs. Firebrands rain down, and radiant heat or flame contact can add to the problem. When the fire on the roof is small, it can be extinguished from the outside. Make sure that they are out; remove the involved shingles to make certain. When fighting exterior roof fires, agency policy must be followed.

When fire has spread across the roof, the structure is seriously threatened, especially in high winds. It must then be assumed that the fire has spread into and through the roof. Only agency personnel trained and equipped for structure fire fighting are permitted to make an interior attack.

Knowing if a roof is too far gone is a judgement call and will depend on your resources, other priorities, etc. Generally, roofs that are more than one-quarter involved are too far gone, and fire fighting resources can be better used to save other structures.

Structure situations to avoid include the following (NWCG):

- Bulging windows and an unventilated roof—hot gases are trapped, and a backdraft is brewing.
- Smoked-over or blackened windows—an interior fire is raging.
- Burning roofs that are 25 percent engulfed in windy conditions—saving the structure is probably hopeless, so stay away.

Water and Foam Use

The wise use of water is critical to the success of structure defense efforts. Water is usually in short supply in these situations. Rural water systems are commonly of low capacity or nonexistent. Even good supplies were not designed to handle dozens of structure fires simultaneously, not to mention ground cover fires. All too often, power failures shut down system pumps anyway.

Water Supply

Conserve water by using only enough to accomplish the task at hand. Save a 100-gallon (400 L) reserve in your engine. This water is for your engine and crew if threatened or need to escape. Take advantage of any opportunity to add water to your tank; that is, if it does not take you out of position at a bad time and does not require an undue amount of time. For example, run a garden hose in your tank while you are parked or stop at a hydrant along your way.

Know the characteristics of the water supply you are relying on, whether it be hydrant, residential supply system, water tenders, engines drafting from open sources, etc. Consider the following:

- System capacity: How much total water is available? When can it be expected to run out?
- Flow rate: How many gallons per minute can you count on? Will that be continuously available, as from pipes or supply engines, or will it be intermittent, as from water tenders?
- Pressure: What will be the pressure at the source you use? Is it adequate to run hoselines directly?
- Reliability: Is the system dependent upon pumps or is it gravity flow? Can water use elsewhere drop your pressure?

Water Application

Effective application is the key to conserving available water. As the ground cover fire approaches, heat begins to build up and firebrands may accompany it. When the fire involves the ground cover fuels around the structure, the heat impinging on you and the structure is at its maximum. After the ground cover fuels burn out, the heat wave will subside. Heavy fuel present may continue to generate heat. The timing of water application with respect to the passage of the heat wave is important. While you must make your own decision on how to apply water based on your situation and experience, the following information may be helpful. If you can simply extinguish the fire, go ahead and do it. If you cannot, put the fire out quickly and directly and consider how to make the best use of your water.

Wetting Down with Water

Wetting down is the application of water to fuel and structures before the fire arrives. Generally, wetting down is done to the roof. Wetting down is usually a waste of time and water. In the face of winds, low humidity, and fire, the wetted surfaces will soon dry out and be susceptible to ignition. Water is more effective if saved to put out ignitions actually occurring on the structure. Foam applications will be explained later.

Reducing the Heat Buildup

Water can be used to reduce or limit the potential buildup of heat by the following:

- Increase the fine fuel moistures in grass or pine needles.
- Knock down the fire in surface fuels where it could spread upward into aerial fuels (such as tree crowns). Under severe burning conditions, fire can still move through the crowns from heat built up elsewhere.
- Prevent it from getting into heavy, troublesome fuels such as woodpiles or brush patches.
- Do not waste water on crown fires, heavy fuels, or fully involved structures. The heat output in these situations far outweighs the ability of water to cool it down.
- Apply water directly to very hot structure surfaces that can help prevent ignition (as evidenced by scorching paint and smoke). Water applied directly is more effective than a "water curtain."
- Do not get water on a hot window — glass will break.

Duration of the Heat Wave

The duration of the intense heat produced by burning ground cover fuels depends upon the fuels involved and on the overall burning conditions. In light fuels, such as grass, the flame front will pass a given point in a minute or so. It will generally move past the structure in no more than a few minutes. In brush, such as chaparral, burnout times are longer and spread rates are often lower than for grass under similar conditions. The fire may take 10 to 15 minutes to move past the structure. Crown fires in timber can generate intense heat that may last a considerable time at any given location. Maintain an escape route and safety zone!

Peak Heat-Wave Tactics

During the peak of the heat and smoke, it is very tempting to spray water at the wall of flame, but it will have no effect and will waste water. To escape the intense radiant heat, use the following guidelines:

- Seek refuge in the shade of something that blocks it.
- Duck around a wall.
- Stay below the roof peak on the sheltered side.
- Take shelter in the structure.
- Wait until you have an opportunity to do some good with your water. Then step out and put it where it counts.
- Use the water when and where you have the advantage and not on fire that is burning at its highest intensity.

Class A Foam

Class A foam is an aggregation of small bubbles created by mechanically injecting air into a foam solution (a mixture of water and foam concentrate) by:

- Air aspirating nozzle systems that can produce wet and fluid foams.
- Compressed air foam systems that can produce wet, fluid, and dry foams.

Properties of Foam

Foam increases the working volume of available water through the expansion of air bubbles and breaks down the surface tension of water for greater penetration of fuel surfaces (makes water wetter). Dense foam can be used to insulate fuels from exposure to flame or smother flames by limiting air supply.

Types of Foam

Each type of foam has different capabilities in fire suppression:

- Wet foam:
  - Flows readily and penetrates rapidly, but drains (dissipates) quickly.
  - Works well for mop-up, wetting down fine fuels to create wet lines to burn out from.
  - Apply the foam line immediately ahead of the ignitors with the foam line width being three times the flame length.
- Fluid foam:
  - Flows readily and drains slower than wet foam.
  - Works well for wet lines in fine fuels.
  - Drains slower than wet foam and provides an insulating barrier. Aerial fuels can also be coated with foam in order to keep a surface fire on the surface.
  - Pretreats structure exposures due to the ability of foam to break down the surface tension of water for greater penetration of moisture in exposed areas. Fluid foam can last up to 30 minutes.
- Dry foam:
  - Coats and adheres well.
  - Wets and drains at a slow rate.
  - Can be used to smother flames in burning material; quick, deep penetration is not needed.
  - Insulates and caps-in moisture on structures or anything it is sprayed on.
  - Can last up to one hour if conditions are right.

Structure Treatment

Foam should be applied to the structure by lofting the foam from a distance. Use the following directions:

- Start on the roof, allowing layers to build up and completely cover all combustible surfaces.
- Cover roofs, eves, outside walls, and any combustibles on the ground adjacent to the structure (**Figure 5.32**). Foam will cling to walls and the roof to provide the insulating barrier needed to protect from heat and flying embers.
- Apply wet foam first for moisture penetration and fluid foam second to cover and help insulate the structure. A final covering of the structure with dry foam will help insulate it even longer.

Figure 5.32 Apply foam to the structure and cover roofs, outside walls, and any combustibles on the ground.

- Foam structures before the fire front hits is especially beneficial where crews will not be able to remain on site to provide protection.
- Apply several coatings of foam on the structure to allow the moisture within the foam to penetrate porous materials. The action of the surfactant in foam lets water that would normally run off penetrate and stay with the fuel. Timing is critical when applying foam to a structure in the path of an approaching wildfire.
- Do not apply foam too early to the structure. It may not have the durability to provide optimum protection.
- Do not begin the foaming operation too late. Firefighters may not get all structures coated, or worse, may not be able to escape an area before the flame front hits.
- Begin treating the structure 10 to 15 minutes before the expected front begins.
- Use Class A foam to coat fuel tanks and LPG containers. Foam clinging to the sides of these tanks will cool them and protect them from direct flame contact. However, Class A foam is not intended to be used on flammable liquid fires.
- Allow firefighters trained in hazardous materials to handle any flammable liquids or other hazardous materials involved with fire. Vacate the area and report it.

Hit and Run Tactics

In some very demanding structure defense situations, such as when numerous structures are threatened and limited suppression resources are available, "hit and run" tactics have proven to be effective. Use the following guidelines:

- Do not become tied down to lengthy supply and attack lines in case you are needed at another structure or moving is necessary if your safety is threatened. If possible, limit the hoselines to 200 (60 m) feet or less.
- Resist attaching supply lines to hydrants. Supply lines may block emergency units, and your escape may be compromised.
- Be aware that hit and run tactics do not require heavy streams and that supply lines are rarely warranted. If you must leave in a hurry, abandon the hose and take only the fittings. Most engines carry enough hose to deploy several working lines.

Retreating and Returning

At times, retreat is necessary due to the intensity of the fire. When the heat becomes, or will become, so bad that your safety is compromised, it is time to pull out using identified escape routes and safety zones. Use the following guidelines:

- Use appropriate colored flagging to mark escape routes and safety zones (hot pink is the universal color designating escape routes).
- Ensure that all personnel are accounted for and maintain communication during retreat. Remain calm. Do not let yourself become excited and careless.
- Use caution and watch for hazards along the escape route.

If escape routes are cut off, take shelter in the structure until it is safe to move out. The structure will not immediately burn down and will offer the best protection against heat and smoke. When the worst of the fire has passed, you may be able to return and save the structure. However, there may be new hazards created by the fire, including:

- Downed power lines.
- Burning snags, which can fall or drop large pieces without warning.
- Debris on the road, including rocks and logs.
- Rolling material that comes off slopes, having been made unstable by the recent burn.
- Hot spots next to the road; smoky conditions.
- Weakened bridges or cattle guards.
- Be aware of the fire situation on any mid-slope roads.

Extinguishment and Follow Up

If time permits, mop up all residual burning materials in the vicinity of the structure to prevent spotting from an ignition source that went unnoticed. Remain at the structure until the homeowner returns. If you must leave for other assignments, ensure that patrol units check the structure at regular intervals. In heavier fuels, structure protection may need to be provided continuously to a single threatened house for an extended period of time.

Firing Operations

Firing operations involve the use of fire to conduct burnouts and backfires. Therefore, it is important to understand the difference between the two.

Burning Out

Burning out is used with direct attack. In direct attack, a fireline is built close to the edge of a fire. Burning out is setting fire inside the fireline to consume fuel between the fireline and the fire. It is generally accepted that operations personnel from crew boss on up have authority to burn out.

Backfiring

Backfiring is an indirect method of attack **(Figure 5.33)**. It is the act of setting fire inside the fireline to:

- Consume the fuel in the path of a fire.
- Change direction or force of the fire's convection column.
- Slow or change the fire's rate of spread.

The Operations Section Chief usually makes the decision for backfiring and is based on recommendations from other operations personnel.

When to Burn Out or Backfire

Make the decision to burn out or backfire when:

Figure 5.33 Backfiring is the act of setting fire inside the fireline.

- You cannot wait for the main fire to reach your established control line.
- The control line will not hold the main fire if it moves against it at full force.
- The intensity of the main fire at the control line would be great enough to threaten the structure.

> **CAUTION:**
> Use of fire is dangerous. Maintain communication and coordination at all times. Make sure that you know your agency's policy concerning burning out and backfiring.

Timing and Coordination

Consider the following:

- Do not perform firing if the fire will create problems for adjoining forces or would result in a threat to other structures in the area.
- Do not initiate firing until the control line to hold it is in place.
- Make sure that firing is necessary. Do not make the decision to fire without consultation with Command and other forces in your area.
- Coordinate your firing operation with those around you and with those planning the overall attack. Let them know of your plan; advise when you begin firing.
- Wait for favorable conditions such as appropriate wind or humidity.
- Make sure you have adequate forces available to patrol the firing operation. Firing to strengthen the control lines should be done as soon as the previous concerns are met.

Control Lines for Firing Operations

Initiate all firing operations from a safe anchor point and, if possible, established control lines.

Mineral Soil:

- Constructed with hand crews or mechanical equipment.
- Make sure that the control line is wide enough to hold the fire.

Natural or Human-Made Features:

- Rock outcrops
- Dirt roads
- Asphalt or agricultural fields

Wet Lines (water/foam/retardant). A wet line is a wetted strip using water, foam, or retardant to act as a control line.

- Where fuels are light, such as grass or litter, use a wet line to control the firing operation. Wet lines are quick and easy to create.
- If the fuel is low and easily penetrated, it works well to wet the strip and then to fire it. The fire goes out as it burns to the wet line.
- If the fuel is high or matted down, the wet line will not penetrate deeply enough. The fire will creep back under the line after the firing operation has moved on. In such cases, light the fire first, then use the water stream to control the inner edge of the fire, making sure it is extinguished.

- When firing from a wet line, it often helps to take advantage of areas of lower fuel. Examples include grass that has been grazed down and tire tracks from vehicles where the grass has been crushed.

Firing and Holding

Personnel assigned to firing and holding operations must be fully briefed and they must also be under the supervision of a qualified firing boss.

Basic Firing Operations:

- In any firing operation, the overall progress along the line should be against the wind or slope that is pushing the fire along the line. In other words, take the firing operation into the wind or down the slope. If wind and slope oppose each other, key on the one that is the strongest.
- If the weather conditions are in your favor, the fire will move quickly away from the control line and should cause no real problems. Just light the edge of the fuel along the control line.
- Fuels outside and adjacent to the control line can be wet down ahead of the firing operation to prevent spotting. Foam works very well for this application.
- Space personnel and equipment out along the line. Do not advance the firing operation until the fire along the line is no longer a threat at that location. The firing operation should not move ahead any faster than the holding operation can keep up with.

Firing Techniques:

There are many firing techniques, but the following are two that work well on interface fires:

- **Strip Firing** — Involves setting fire to one or more strips of fuel and allowing the strips to burn together. Lighting numerous strips allows faster area ignition. By varying the width of the strips and their location in relation to the slope or wind direction, you provide a means of regulating the fire's intensity.
- **Ring Firing** — A technique generally used as an indirect attack and backfire operation. It involves circling the perimeter of an area with a control line and then firing the entire perimeter. Ring firing is often used to burn out around structures. However, firing personnel may not have a strong anchor point to commence firing. Escape routes and safety zones must be established.

Holding Firing Operations:

- Deploy engines, hoselines, or hand crews along the line behind the firing operation.
- Make sure that the holding operation is capable of dealing with hot spots or escaped fire across the control line.
- Do not impair the intentional fire. Knock down hot spots and flare-ups that threaten to escape, either by flame or firebrands.
- Put all the necessary resources to work to contain an escape if one occurs. Advise the crews doing the lighting of the escape so that they can slow down or stop until the escape is controlled.

Follow-Up

Once the fire front or major heat wave has passed your position, your job is not yet completed. Nothing would be more frustrating than defending a structure from the heat, smoke, and flames of the fire front or leaving to assist another company or crew and returning only to find the first structure totally consumed from a hidden spark. Do not let your desire to move with the fire front overpower your obligation to finish the job at hand. Your initial concern should be the structures you were assigned to protect.

Check the structure for fire at likely ignition points. You must check for sparks or embers at:

- Roofs
- Siding
- Under eaves and in rain gutters
- Vents
- Under decks and porches
- Wood piles

Check for heat or flame extension into the interior:

- Attics
- Curtains or windows
- Furniture
- Carpets
- Wall
- Cupboards
- Ducts

Perform only enough mop-up or overhaul to ensure structure safety before moving to other structures. Complete extinguishment of any fire in or on the structure. Provide a positive barrier between a surface fire and the structure. If the owners are present, instruct them as to what they can do to continue protection and mop-up (remember the risks involved). Leave all homeowner's ladders and garden hoses in place and ready to use.

When a site becomes secure and equipment becomes available, contact Command for your next assignment. If possible, always try to leave one engine to patrol the burned area and to assist with mop-up and security.

Before Leaving the Area

- Provide for patrol by engines or crews.
- Leave a note on the door or entry telling occupants:
    — What you did with utilities
    — What happened to pets, if applicable
    — Who entered the home and why
    — Date and time
    — Your signature and title
    — Your business card

- Leave on a few lights so that patrol crews can locate the structure.
- Secure the structure. Be sure to brief the patrol crews that their primary mission is to provide structure protection.

Patrol Duties

To prevent further losses, it is important to remember that the primary responsibility is to structures and to minimize damage to improvements and the environment. This includes checking the following:

- Water system
- Fences
- Damaged buildings
- Security plans

Your duties also include:

- Assist and instruct homeowners as to effective mop-up procedures.
- Maintain a high visibility to the homeowner. This is critical before, during, and after the fire.
    — Homeowners who stayed or are returning want to see a fire engine.
    — Practice positive public relations in an attempt to leave the public with positive feelings. Example: Leave a crew that worked in the area that has some emotional ties to the work accomplished.
- Document damage.
    — Document noticeable damage
        - Take a picture.
        - Shift plan information.
    — Keep an accurate record of damage caused by fire suppression action.
        - Count the number of damaged structures.
            ▷ Homes
            ▷ Outbuildings
            ▷ Improvements
            ▷ Vehicles/Equipment
            ▷ Crops
            ▷ Order a fire investigator or claims specialist as needed

Public Relations

Seldom will the need for effective public relations be greater than during and after an interface fire. However, it is critical that public relations programs need to be operational before the fire **(Figures 5.34 a and b, p. 148)**. Making the public aware of the problems that exist before there is a fire can help develop the cooperation needed when a fire occurs. It is also prudent to identify those media contacts who will be there to cover these events. Identify these contacts before the fire season and invite them to planning or preincident surveys or even have them take basic fire training.

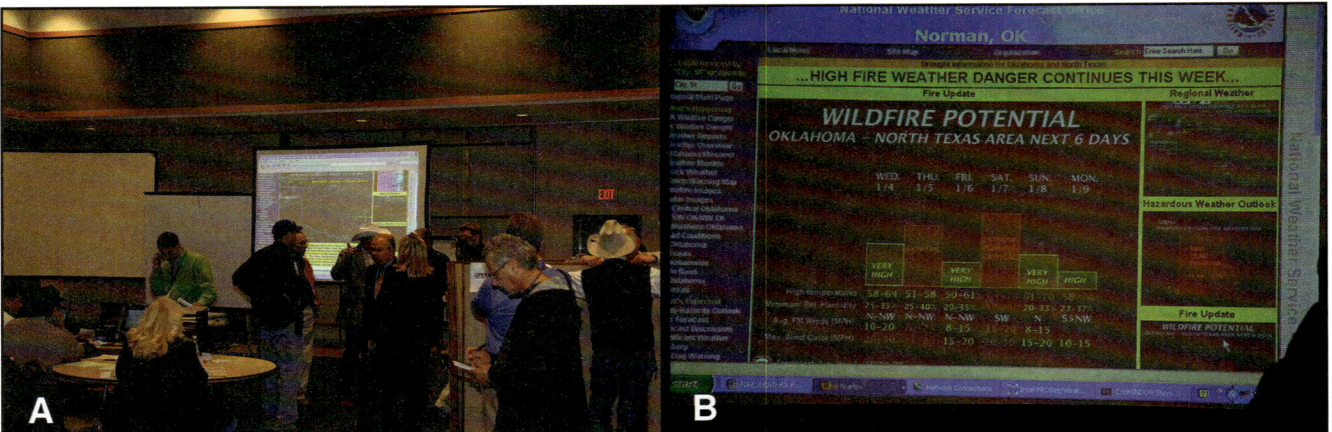

Figures 5.34 a and b Before fires occur, a public relations program needs to be operational.

Homeowners threatened by wildfire will be experiencing a great variety of emotions: fear, apprehension, anger, etc. Many of these emotions may be directed toward the firefighters. As an Incident Commander or Company Officer, you will be required to control your emotions while trying to maintain control of the incident at hand. Expect panicked homeowners and remember your mission.

There may not be a Public Information Officer present. Therefore, as an Incident Commander or Company Officer, be prepared for a multitude of questions from people who have no idea of what is going on. Be prepared to interact with the resident or residents. Questions may include the following:

- Where is the fire?
- Where is it going?
- What started it?
- Are you evacuating?
- Has anyone been hurt; have homes burned?
- Why didn't you put this fire out when it started?
- Why aren't you in there protecting homes?

Request a Public Information Officer or a member of the Public Information Office staff as soon as possible. Follow the guidelines of the Public Information Officer for the release of information. If in doubt, find out or say something positive, such as, "We are doing the very best job we can." Those in leadership positions must be prepared to run interference between the public and their crews.

Dealing with the Media

Nothing brings out the news media faster than a disaster. Expect large numbers of media representatives wandering around the fireground. When facing the media:

- Be courteous and act professional, but do not let them interfere with your job.
- Refer their questions to a Public Information Officer or Command.

- Do not provide information that is undocumented or the names of those injured or killed.
- Remember that radio communications may be monitored by the media and the public.

Dealing with the Public

Public opinion of firefighters will remain long after the interface fire is out. Maintain a professional attitude, but be sensitive to the needs of the affected public. Be careful of what you say and how your crew acts at all times. Consider the following guidelines:

- Always try to minimize damage caused by control methods.
- Document any damage caused directly by suppression actions using a disposable or digital camera.
- When time permits, talk to the homeowners and explain what actions were taken and why. Something that can leave a lasting positive impression with property owners is to follow good salvage operation practices.
- People often have a greater emotional attachment to pets than other personal property.
- If your agency policy authorizes entering structures:
  — Be cautious if you do enter the structure.
  — Ensure the protection of valuables and heirlooms.
  — Place smaller items into closets and close doors.
  — Ensure the safety of photos and pictures, video equipment, computers and other high value items.
- Remember this cardinal rule – **Always treat the property of others better than how you would want your own property treated.**

Key Terms

Backfire — Fire set along the inner edge of a control line to consume the fuel in the path of a ground cover fire and/or change the direction of force of the fire's convection column, (National Wildfire Coordinating Group (NWCG) *Glossary of Wildland Fire Terminology*).

Backfiring — A tactic associated with indirect attack, intentionally setting fire to fuels inside the control line to slow, knock down, or contain a rapidly spreading fire. Backfiring provides a wide defense perimeter and may be further employed to change the force of the convection column. Backfiring makes possible a strategy of locating control lines at places where the fire can be fought on the firefighter's terms. Except for rare circumstance meeting specified criteria, backfiring is executed on a command decision made through line channels of authority. (National Wildfire Coordinating Group (NWCG) *Glossary of Wildland Fire Terminology*).

Booster Line — Noncollapsible rubber-covered, rubber-lined hose usually wound on a reel and mounted somewhere on an engine or water tender. This hose is most commonly found in ½-, ¾-, and 1-inch (13 mm, 19 mm, and 25 mm) diameters and is used for extinguishing low-intensity fires and mop-up. *Also called* Hard Line.

Burning Out — Setting fire inside a control line to consume fuel located between the edge of the fire and the control line. (National Wildfire Coordinating Group (NWCG) *Glossary of Wildland Fire Terminology*).

Evacuation — An organized, phased, and supervised withdrawal, dispersal, or removal of civilians from dangerous or potentially dangerous areas, and their reception and care in safe areas. (National Wildfire Coordinating Group (NWCG) *Glossary of Wildland Fire Terminology*).

Flanking Fire Suppression — Attacking a fire by working along the flanks either simultaneously or successively from a less active or anchor point and endeavoring to connect two lines at the head. (National Wildfire Coordinating Group (NWCG) *Glossary of Wildland Fire Terminology*).

Hand Crew – A number of individuals who have been organized and trained and are supervised principally for operational assignments on an incident. (National Wildfire Coordinating Group (NWCG) *Glossary of Wildland Fire Terminology*).

Hose Lay — (1) Arrangement of connected lengths of fire hose and accessories on the ground at a ground cover fire beginning at the first pumping unit and ending at the point of water delivery. (2) Connected lengths of hose from water source to pumping engine.

Hotshot Crew — Intensively trained fire crew used primarily in hand line construction (Type-1). (National Wildfire Coordinating Group (NWCG) *Glossary of Wildland Fire Terminology*).

Hot-spotting — Checking the spread of fire at points of more rapid spread or special threat. Is usually the initial step in prompt control, with emphasis on first priorities. (National Wildfire Coordinating Group (NWCG) *Glossary of Wildland Fire Terminology*).

Mobile Attack — Suppressing fire along a fire edge by driving mobile apparatus along the perimeter and simultaneously applying fire streams to knock down the fire. *Also called* Pump and Roll.

Patrol — (1) To travel over a given route to prevent, detect, and suppress fires. (2) To go back and forth vigilantly over a length of control line during and/or after construction to prevent slopovers, suppress spot fires, and extinguish overlooked hot spots.

Pincer Attack — Direct attack around a fire in opposite directions by two or more attack units with the ultimate intent of pinching off (stopping) the head of the fire.

Shelter in Place — Remaining in a structure or vehicle when a fire moves through rather than attempting to use roads that may be blocked or untenable because of fire; opposite of evacuation.

Structure Triage — Process of inspecting and classifying structures according to their *defensibility/indefensibility* based on their situation, their construction, and the immediately adjacent fuels.

References

National Wildfire Coordinating Group (NWCG) *Glossary of Wildland Fire Terminology*. Accessed online. https://www.nwcg.gov/glossary/a-z

NIMS Resource Management & Mutual Aid. Accessed online. https://www.fema.gov/resource-management-mutual-aid

Wildland Fire Incident Management Field Guide, National Wildfire Coordinating Group, PMS 210, (2013). Accessed online.

Wildland Fire: U.S. National Park Service. Accessed online.

Chapter 6

Water Supplies and Support

Table of Contents

Water Sources 155
    Piped Systems 155
    Rural Water Sources...................................157

Mobile Water Operations 158
    Safe Operation, Driving, Off-Road Use 158

Portable Water Tanks, Types, Sizes, Proper Location 159

Water Shuttle Operations......................................159

Nurse Tender Operations 160

Pumping and Dumping Capabilities 160

Tactical Tender/Tanker Operations 161

Water Supplies and Support

Chapter 6

The agent most often used to extinguish ground cover fires is water. For water to be effective as an extinguishing agent, it must be available in sufficient quantity. Ground cover fires often occur in areas where a public water supply system is not available. Therefore, firefighters must take advantage of auxiliary water sources near the fire or depend on water (and other agents) being transported to the fire. Having access to water sources and the efficient use of that water can significantly affect the outcome of efforts to control ground cover fires. Firefighters must know the location of water sources within their response districts and be prepared to take full advantage of them.

Water Sources

An adequate water supply and distribution system is an essential tool firefighters need to control and extinguish fire. Water is easily stored and can be transferred over large distances in a well-designed water distribution system.

Piped Systems

A piped (municipal) water supply system may be available to the woodland, brushland, and grassland areas in or near cities and towns. These public or private systems consist of water mains (pipes) that carry the water from the source to the point of use — individual homes, businesses, or fire hydrants and standpipes **(Figure 6.1)**. Most modern water systems now use at least 8-inch (200 mm) diameter pipe for mains. In older municipal systems, and some rural systems, smaller mains may still be in use. It is important that firefighters become familiar with the capabilities and limitations of any public or private water systems in their response districts *before* these systems are needed for fire fighting. Of most importance are the locations of hydrants, the rate of flow to be expected from the hydrants, and the total volume of water available.

Figure 6.1 Fire hydrants are a fire service connection point in public and private water systems.

Figure 6.2 A common practice is to paint standpipes and privately owned hydrants red.

Another aspect of piped water systems with which firefighters should become familiar is the color-coding system used by the local water utility to mark hydrants and standpipes. While local entities are free to use whatever colors they choose, the most common practice is to paint standpipes and privately owned hydrants all red (**Figure 6.2**). The barrels of all public hydrants are usually painted a highly visible color (white, yellow, etc.). However, the caps and bonnets of individual public hydrants are usually painted a contrasting color to indicate the flow in gpm (L/min) that can be expected from that hydrant (**Table 6.1**). Some entities also use blue on the caps and bonnets of hydrants, such as those downslope from a storage tank, that have very high static pressure (**Figure 6.3**).

Figure 6.3 In some areas, the top of a hydrant may be painted blue indicating that the hydrant has high static pressure.

Table 6.1
Classifications and Markings of Municipal Fire Hydrants

Classification	Fire Flow	Barrel Color	Top and Nozzle Cap Colors	Pressure
Class AA	1,500 gpm (5 680 L/min) or greater	Chrome Yellow	Light Blue	20 psi (140 kPa)
Class A	1,000–1,499 gpm (3 785–5 675 L/min)	Chrome Yellow	Green	20 psi (140 kPa)
Class B	500–999 gpm (1 900–3 780 L/min)	Chrome Yellow	Orange	20 psi (140 kPa)
Class C	500 gpm (1 900 L/min) or less	Chrome Yellow	Red	20 psi (140 kPa)

Based on information given in NFPA® 291, Recommended Practice for Fire Flow Testing and Marking of Hydrants, 2012.

Hydrants should be tested at least once each year to ensure that they are operating and that all parts are in good working order. Even though these systems can supply large quantities of water for fire fighting purposes, the water should never be wasted. A well or surface reservoir with sufficient capacity for normal domestic consumption may supply a small water system but may be incapable of supplying the additional quantities needed for fire fighting. If electrical power to the water pumps is lost for any reason, the volume of water available from the system may be greatly reduced. Excessive domestic water use from piped water systems can also reduce the volume and pressure available for fire fighting operations. In addition, if a number of engines are connected to hydrants and all are pumping at or near capacity, the residual pressure in the water system may drop significantly. This can lead to vacuum pockets forming due to localized regions of low pressure at the vanes in the impeller of a centrifugal pump causing vibrations, loss of efficiency, and possibly damage to the impeller (known as *cavitation*) in the fire pump. This may lead to possible water main collapse or contamination as the result of backflow intrusion into the domestic water system.

Rural Water Sources

Water is supplied from natural freshwater sources such as wells, springs, rivers, lakes, and ponds **(Figure 6.4)**. Streams and springs may be fed by groundwater or snow-melt at higher elevations. Rainfall replenishes lakes, ponds, and groundwater. However, it is sometimes difficult to position fire apparatus close enough to these sources to use them. During preincident planning, identify likely sites where streams can be dammed during a fire. Encourage residents who have one or more auxiliary water sources on their property to construct suitable approaches for fire apparatus. A suitable approach is one that will support the weight of a fully loaded fire engine. The value of these auxiliary water sources is significantly enhanced with the addition of a dry hydrant connected to a pipe that extends under the water **(Figure 6.5)**.

Figure 6.4 Water for community use is supplied from natural freshwater sources such as lakes and ponds.

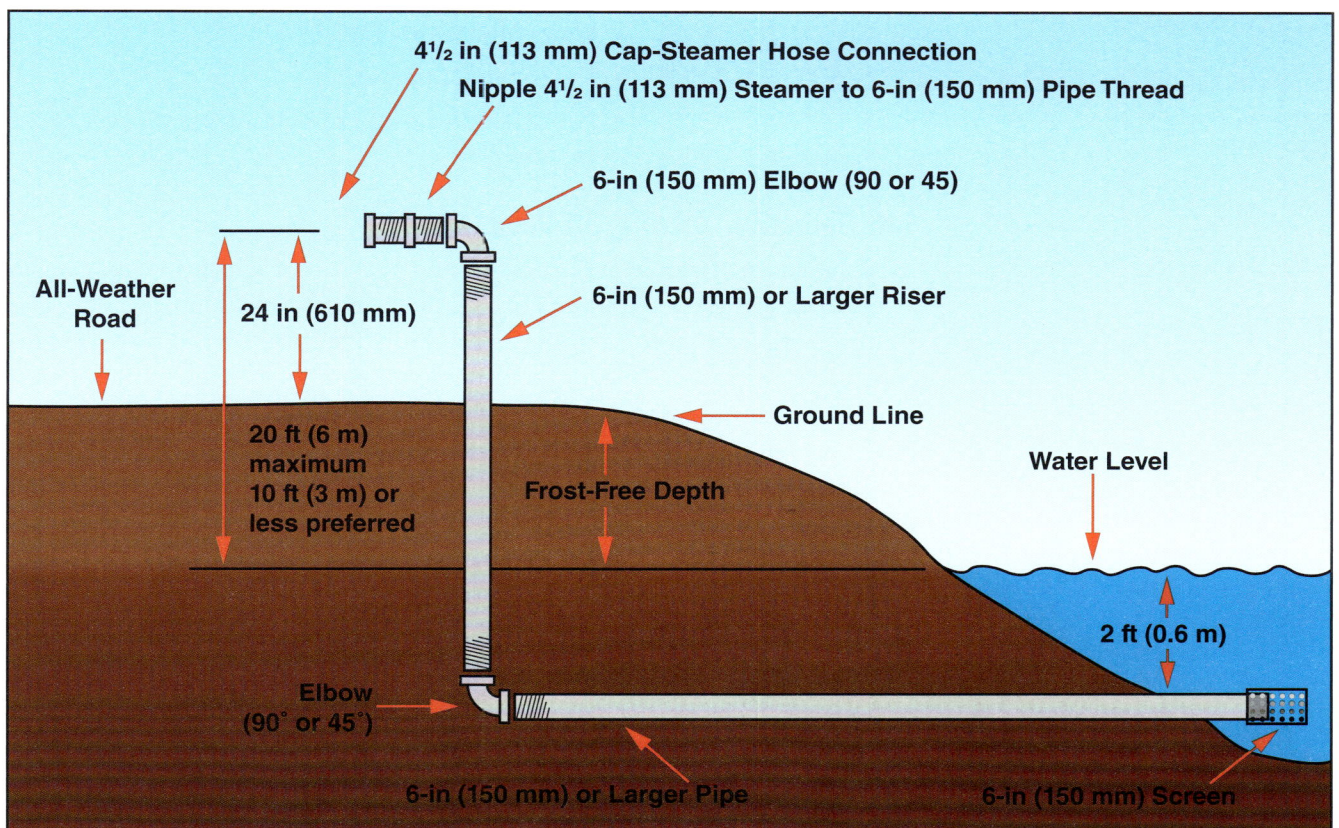

Figure 6.5 A typical dry-hydrant installation.

Figure 6.6 Mobile water supply apparatus are widely used to transport water to areas without piped systems or auxiliary water sources.

Figure 6.7 Construction water trucks may be used for water shuttle operations.

Mobile Water Operations

Mobile water supply apparatus (water tenders/ground tankers) are widely used to transport water to areas without piped systems or auxiliary water sources (**Figure 6.6**). In an emergency, almost any vehicle with a large-capacity tank can be used for this purpose. Some of the auxiliary water-carrying vehicles that may be used are street flushers, sprinkler trucks, ready-mix concrete trucks, milk trucks, water trucks from building contractors, highway maintenance equipment, and even railroad tank cars (**Figure 6.7**). Auxiliary water-carrying vehicles may be rented or borrowed during emergencies under prior agreements made with their owners during the preincident planning process.

As with ground cover engines and other resources, water tenders/ground tankers are classified into various types based on their *minimum* capabilities. In general, the three types of water tenders have pumps ranging from 50 to 300 gpm (200 L/min to 1 200 L/min) and water tank capacity ranging from 1,000 to 5,000 gallons (4 000 L to 20 000 L). Depending on local operating procedures and the capabilities of particular apparatus, three basic fireground applications are available for water tenders.

- Water shuttle operations
- Nurse tender operations
- Fire attack/exposure protection operations (tactical tenders)

Safe Operation, Driving, Off-Road Use

Extreme caution is necessary in driving and operating a vehicle that may weigh as much as 60,000 pounds (27 215 kg). Stopping distances are extreme, and shifting water can cause handling and stability issues on curved roads and inclines. Anyone responsible for the driving or operation of these vehicles must be qualified through a standardized program such as Fire Apparatus Engineer (FAE) or Vehicle/Machinery Operations (VMO). Use extreme caution when operating and supporting suppression operations off road. Vehicle weights and the nature of the design of the vehicle, with such a high center of gravity, can be dangerous on side slopes, unstable landscapes, and in areas with limited maneuverability. Common sense should prevail. If you have any question as to whether you should proceed, you probably should not. Wear seat belts at all times, and always use your spotters when positioning any piece of fire apparatus.

> **NOTE:** All fire personnel tasked with vehicle operations has a responsibility to know and practice the standard operating procedures for their respective jurisdictions and/or statutory requirements set by the state in which they reside and respond.

Figure 6.8 Left-Square metal-framed and lined portable tank; Right-Round self-supporting portable tank.

Portable Water Tanks, Types, Sizes, Proper Location

With some special exceptions, there are two types of portable water tanks. One is the collapsible or folding style that uses a square metal frame and a synthetic or canvas duck liner. Another style is a round, self-supporting synthetic tank with a floating collar that rises as the tank is filled (**Figure 6.8**). These frameless portable tanks are widely used in ground cover fire fighting operations.

Whatever the size available is for the operation at hand, the key is proper placement. As every jurisdiction has established operating standards and guidelines, some general observations are noted:

- Maintain flat/level surface.
- Do not block access/egress routes/entrances.
- Provide easy approach for support apparatus.
- Do not position under power lines.
- Provide protection from approaching fire front and ongoing burnout operations.

Water Shuttle Operations

Water shuttle operations involve the more-or-less constant movement of one or more water tenders between a water source (the fill site) and the location where the water is going to be used (the dump site) (**Figure 6.9**). The fill site is usually a fire hydrant or a static water supply source. Typically, an engine is standing by at the water supply source ready to quickly fill the water tenders as they arrive.

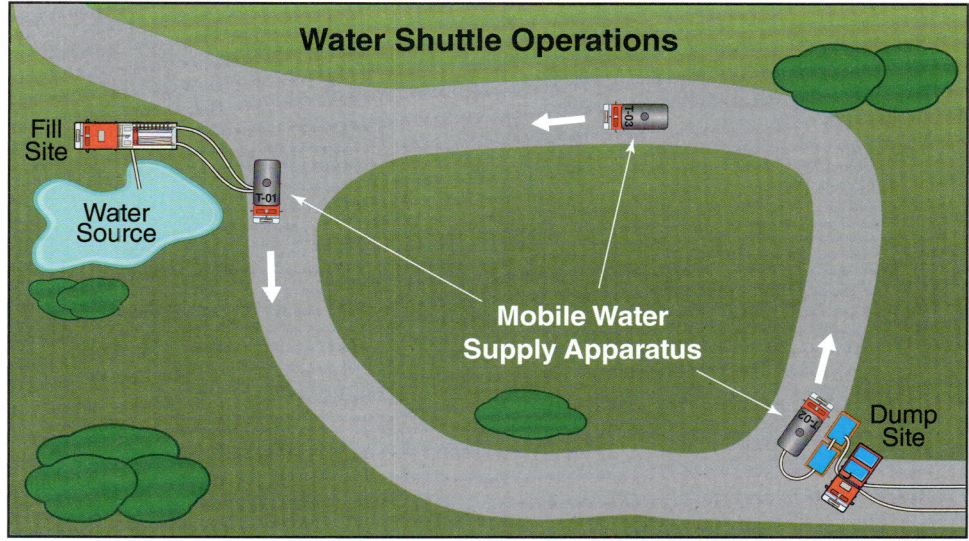

Figure 6.9 Water shuttle operations.

Once at the dump site, water tenders dump their loads into one or more portable water tanks and immediately return to the fill site for another load. The key to timely and successful water shuttle operations is good communication between all parties involved and a safe, direct route between fill and dump sites. Speed is not always the most important element. The right number of shuttle apparatus in motion at a constant rate maintains a steady supply faster than a few vehicles operating at a fast pace.

Nurse Tender Operations

Nurse tender operations involve pairing a water tender/ground tanker with an engine to extend the engine's water supply. For example, a Type-1 engine may be assigned to protect a structure from an approaching fire or supply a progressive hose lay attacking a hot flank. If the amount of water in the engine's tank is not sufficient to allow it to complete its assignment, the water tender/ground tanker connects a supply line to the engine. When the situation requires that the water tender/ground tanker be positioned some distance from the engine, it can be connected by a long supply hose if the water tender/ground tanker has an adequate fire pump. A water tender/ground tanker used as a refilling point for smaller ground cover apparatus operating remotely from a water supply source also qualifies as a nurse tender operation.

Water tenders/ground tankers equipped with suitable fire pumps may be able to make fire attacks or perform exposure protection by themselves. In order to do these functions, the fire pump needs to be capable of providing adequate volume and pressure for attack lines. Under some fire conditions, water tenders/ground tankers are staged in an area of potential exposures to provide protection for that area while being available to refill ground cover apparatus when required.

Pumping and Dumping Capabilities

Three primary methods are used to get water out of water tenders/ground tankers:

- Pump the water through the fire pump.
- Use a pressurized dump of the water through a quick-dump discharge.
- Unload the water by gravity through a gated dump valve.

Pumping through the apparatus fire pump is generally done when water tenders/ground tankers are supplying attack lines or acting as a nurse tanker. The types of fire pumps carried on mobile water supply apparatus range from small auxiliary engine-driven pumps to midship transfer drive pumps. However, pumping water from the apparatus is generally not the most efficient method of filling portable water tanks during water shuttle operations. The most efficient way to discharge water into a portable tank is through the use of a quick-dump discharge system **(Figure 6.10)**. The quick-dump system connects directly from the apparatus water tank, through the apparatus body (if there is any), to the sides and/or rear of the apparatus. The quick-dump piping may be round or square and is generally at least 8-inches (200 mm) in diameter. The quick-dump system may rely on gravity to unload the water, or it may be jet-assisted by a small water line from the fire pump. The quick-dump discharge valves may be located on the dump itself, or they may be remotely controlled from the apparatus cab.

Figure 6.10 A quick-dump discharge system.

Some water tenders/ground tankers are not equipped with any type of fire pump. These apparatus are generally limited to gravity dumping their loads in water shuttle operations. If these apparatus have the appropriate connections, engines may be able to connect to them with hard suction hose and draft the water from the water tender's/ground tanker's tank.

Tactical Tender/Tanker Operations

The ability to pump while the vehicle is in motion is particularly useful for ground cover fire apparatus. The speed with which ground cover fires spread frequently requires firefighters to attack while on the move. Pump-and-roll operations are very effective when making a direct attack on low-intensity ground cover fires in terrain that is suitable for the type of apparatus. Apparatus equipped for pump and roll with Class A foam are also well suited for laying a protective foam blanket over fuel and structures exposed to a ground cover fire. Most pump-and-roll capable apparatus use either a power take-off or an auxiliary engine pump. Water is discharged during pump-and-roll operations by any one or more of the following means:

- Booster reel
- Short section of adequate handline-sized hose
- Ground sweep nozzles
- Remote control nozzles
- Heavy stream nozzles

Booster reel lines are sometimes used for pump-and-roll operations because the length of the hose off the apparatus can be adjusted easily. Booster hose is also very durable — a necessity for ground cover operations where the hose is subjected to much abuse. Booster hoses vary in diameter, and the hose may be either the low- or high-pressure type, depending on the types of fires that are anticipated.

Some departments prefer to equip the apparatus with a shortened section of fire hose connected directly to a small discharge located on the front, rear, or sides of the apparatus. These shortened sections of hose allow firefighters to easily apply water or foam while walking alongside a slowly moving apparatus. However, in flashy fuels with high heat output or long flame lengths, 1 ½ -inch (38 mm) hose must be used for safety. The relatively high friction loss in ¾ - or 1-inch (20 or 25 mm) booster hose limits the quantity of water available from these lines. Booster lines must not be used when the thermal output of the fire exceeds the heat-absorbing capacity of the water being discharged.

Ground sweep nozzles:

- Used when the fuel is relatively short and the fire intensity is low.
- Mounted either at both corners of the bumper or as a single spray bar that spans the entire width of the bumper.
- Controlled from inside the cab.
- Mounted at the front or rear of a vehicle

Remote control nozzles:

- Used for pump-and-roll operations.
- Mounted above the front bumper
- Controlled from within the cab of the apparatus.
- Capable of flows of 10 to 300 gpm (40 to 1 200 L/min)
- Operated in the low end of that flow range when attacking ground cover fires to avoid wasting water.

Incident Commanders and line officers alike must keep in mind that tactical tenders/ground tankers have limited maneuverability and are slow moving and very heavy. The practicality of remote off-road use is limited. Tactical tender/ground tanker use is best utilized in wildland urban interface operations where large volumes of water and limited resources make it an excellent asset.

Chapter 7

Tactical Resource Support and Operations

Photo courtesy of NIFC.

Table of Contents

Fire Crews **165**
    Interagency Hotshot Crews 166
    Hand Crew Duties and Responsibilities 166

Heavy Equipment **167**
    Dozers 168
    Tractor-Plows 168
    Road Graders and Other Mechanized Equipment 169

Hand Tools **170**
    Cutting Tools 170
    Scraping Tools 171

    Fire Swatters (Flails) 172
    Wire Brooms 173
    Backpack Pumps 173

Fire and Aviation Management **174**
    Helicopter Operations Safety Overview 174
    Helicopters 175
    Fixed-Wing Aircraft 176
    Smokejumpers 176
    Single-Engine Air Tankers (SEAT) 177

References **178**

Tactical Resource Support and Operations

In order to successfully combat ground cover fires, any number or combination of human and technical resources may be required. This includes hand crews, heavy equipment, hand tools and equipment, and aviation resources. This chapter reviews each of these resources.

To manage a fire, it takes a variety of people who each have a different skill set. This includes firefighters who have diverse skills as well as those behind the scene who support the firefighters in their efforts. With proper organization, each firefighter will be individually prepared and upon arrival at a fire have a specific task to perform.

A number of different types of crews may be involved in ground cover fire fighting, and they all need to be ready to perform at their best when called. To work safely and effectively, firefighters need to be:

- Well-trained
- Properly equipped
- Physically fit
- Well-fed and hydrated
- Given reasonable rest periods between work cycles as required by state/provincial and federal regulations.

Fire Crews

Fire crews are the backbone of the fire suppression effort. According to the National Park Service, crews are differentiated between Type 1, Type 2, and Type 3 crews based on experience, leadership and availability. Each crew is comprised of between 18-20 men and women. A crew boss and three squad bosses supervise the actual work the crew does.

Fire crews are on the frontlines using either direct or indirect suppression tactics in battling wildfires. For smaller fires, direct attack involves cutting fireline around the fire or putting out flames with water or soil. On larger or rapidly moving fires, indirect attack is used. This involves using natural firebreaks such as roads, trails or streams as the control line and removing the fuels between it and the fire by backburning.

Interagency Hotshot Crews

The most experienced Type 1 crews are known as Interagency Hotshot Crews (IHCs) **(Figure 7.1)**. IHCs are diverse teams of career and temporary agency employees who uphold a tradition of excellence and have solid reputations as multi-skilled professional firefighters. Crews are employed by the USDA Forest Service, Bureau of Land Management, National Park Service, and Bureau of Indian Affairs and Tribal programs. Their physical fitness standards, training requirements, and operation procedures are consistent nationwide, as outlined in the *Interagency Hotshot Operations Guide*. Core values of "duty, integrity, and respect" have earned Hotshot crews an excellent reputation throughout the United States and Canada as elite teams of professional wildland firefighters. Hotshot crews started in Southern California in the late 1940s on the Cleveland and Angeles National Forests. The name was in reference to being in the hottest part of fires (National Park Service).

Figure 7.1 A Type I Interagency Hotshot Crew — the 2017 Entiat Hotshots. *Courtesy of the US Department of Agriculture, US Forest Service, Pacific Northwest Region (R-6).*

Hand Crew Duties and Responsibilities

Hand crews serve as the infantry of ground cover fire forces. According to the USDA Forest Service, the crew's main responsibilities are to construct a "fireline" around ground cover fires and wildfires. The USDA Forest Service provides the following guidelines for crew members.

- To construct a fireline from the anchor point:
    - Clear a strip of land from flammable materials.
    - Dig down to the mineral soil to control fires.
    - Burn out fire areas.
    - Mop up after the fire.
- Help construct containment lines.

- Cut open smoldering trees to put out the sparks and spread water dropped by helicopters or supplied by engine crews. Water is not always available; at those times, firefighters move everything from a hot spot, separating burning limbs and coals and covering them with dirt. They also patrol firelines "to make sure nothing creeps over into other areas."

> **NOTE:** It is critical to firefighter safety to start fireline construction from an anchor point. An anchor point can be a road, a lake, a stream or river, or a large rock outcropping. It is used to prevent the fire from flanking the crew as they construct line around the perimeter of the fire.

The general duties of a hand crew:

1. Perform manual and semi-skilled labor.
2. Ensure that objectives and instructions are understood.
3. Perform work in a safe manner.
4. Maintain self in the physical condition required to perform the arduous duties of fire suppression.
5. Keep personal clothing and equipment in serviceable condition.
6. Report to supervisor any close calls, accidents, or injuries.
7. Report hazardous conditions to supervisor.

The National Wildfire Coordinating Group (NWCG) sets the following standards as to how much fireline a type of hand crew should construct in an hour. Lengths are noted by numbers of chains. A chain equals 66 feet (20 m).

- A Type 1 crew is expected to complete 30 chains (1,980 feet [604 m]) of line in short grass per hour.
- A Type 2 crew should complete 18 chains (1,188 feet [364 m]) in an hour.
- In brush, a Type 1 crew should complete 6 chains (396 feet [120 m]) in an hour.
- A Type 2 crews should complete 4 chains (264 feet [80 m]) in an hour. The fireline has to be taken down to mineral soil with no combustibles inside it.

In short grasses and light fuels, a 12- to 36-inch (300 mm to 1 m) fireline should halt the advance of the fire. In denser and larger fuels, a fireline is merely a starting point. It may take an area of fireline, void of combustibles, more than 100 feet (30 m) wide to halt the fire's spread. To achieve this, firefighters often rely on burnouts. If conditions permit, crews will start a fire at the fireline and allow it to burn to the approaching main fire. With all of the combustibles in its path depleted, the fire should die.

Heavy Equipment

In addition to standard fire fighting apparatus, different types of heavy equipment may be used to support a major ground cover fire fighting operation. This equipment is used primarily to construct firelines that may slow or halt the spread of an advancing fire or to provide a means of anchoring a firing operation. Therefore, firefighters must be familiar with the various types of equipment that are available to them and know their capabilities.

In many cases, heavy equipment used at ground cover fires is not owned or operated by a fire department. It is contracted from other public agencies or private firms. During preincident planning, personnel should conduct a survey of all earth-moving equipment within the jurisdiction to identify the capabilities, limitations, and availability of each type. Written and signed agreements should specify the circumstances under which such equipment may be used and how the owner will be compensated. These agreements should clearly identify the rights and obligations of both parties. If the owner is to provide operators of this equipment during suppression operations, ensure that ALL firefighter certifications, such as a NWCG Red Card, NFPA® 1001 certification for firefighters, or NFPA® 1002 certification for fire apparatus driver/operators, are clear and specified prior to the agreement. Fire department personnel should inspect contract equipment on a regular basis (at least annually) and review the equipment operators' qualifications to operate at ground cover fires. It is critically important that contract operators be trained in safety and ground cover fire behavior — and that this training be maintained through refresher courses prior to the start of each fire season.

The following are the three most common types of mechanized equipment used in ground cover fire fighting and are explained in this section:

- Dozers
- Tractor-plows
- Road graders (maintainers)

Depending on local conditions, other types of equipment may be used as well. Farm plows, large mowers, and similar types of equipment may be used to cut firebreaks if the fire is in terrain, such as crop lands, pastures, and parks, that is suitable for these implements. In many cases, their use is similar to the principles described for the other heavy equipment covered in this section.

Dozers

Three types of dozers are commonly used in ground cover fire fighting operations. Each type has a crew of two.

- Type-1   Heavy Dozer   Minimum 200 HP
- Type-2   Medium Dozer   Minimum 100 HP
- Type-3   Light Dozer   Minimum 50 HP

While there are other uses, the primary use for dozers is the construction of firelines. These man-made barriers are constructed in an attempt to slow the spread of a ground cover fire or serve as a control line from which firing operations may be conducted. Because dozers are capable of operating in a wide variety of fuels, topography, and soil conditions, they are very well-suited to this application **(Figure 7.2)**.

Tractor-Plows

According to the NWCG, the following are the six types of tractor-plows used in ground cover fire fighting operations:

- NWCG Type 1 — Heavy tractor-plow and largest capacity tractor-plow

Figure 7.2 Dozers are commonly used in ground cover fire fighting operations.

- NWCG Types 2 and 3 — Medium capacity tractor-plow
- NWCG Types 4, 5, and 6 — Light tractor-plow; smaller capacity

A *tractor-plow* is a vehicle used to cut a control line in somewhat the same manner as a dozer. However, most tractor-plows are smaller and more maneuverable than dozers. Tractor-plows may be tracked vehicles or rubber-tired vehicles. They pull a plow that is typically about 6 feet (2 m) wide. This cuts a narrower swath than the blade on most dozers.

Road Graders and Other Mechanized Equipment

Road graders, also called *maintainers*, are usually self-propelled and have the blade mounted just forward of the rear wheels **(Figure 7.3)**. Road graders are used for building roads and are not designed to be fire fighting equipment. However, they can be effective in cutting a control line well ahead of a slow-moving fire in crop stubble. To be done safely, the line must be located far enough ahead of the fire front so that it can be completed well before the fire reaches the line.

Figure 7.3 Road graders are effective in cutting a control line.

Depending on local conditions, other types of equipment may be used as well. Farm plows, large mowers, and similar types of equipment may be used to cut firebreaks if the fire is in terrain, such as crop lands, pastures, and parks, that is suitable for these implements. In many cases, their use is similar to the principles described for the other heavy equipment covered in this section.

> **CAUTION:**
> **Trained fire personnel who can communicate and coordinate the use of these machines should supervise operators. These operators generally are not fire qualified and need a thorough briefing on the dangers that exist in the fire environment.**

Hand Tools

Although mechanized apparatus and equipment are widely used, hand tools also have an important place in fighting ground cover fires. Some of the hand tools are conventional, and some are adaptations of conventional tools. Others, however, have been specially developed for fighting fires in ground cover fuels. The selection and design of hand tools usually depend upon the situations where they are likely to be used and upon local preference. Generally, grass fires require more use of scraping or smothering tools, while brush fires require more cutting tools.

When carrying and working with hand tools, firefighters typically walk and work 10 feet (3 m) apart for safety **(Figure 7.4)**. Firefighters should carry tools in the following manner:

- Hold the tool at its balance point.
- Do not run with hand tools.
- Keep the tool at your side and close to the body and not on the shoulder.
- Situate the tool on the downhill side when walking across a slope.
- Signal and wait for the right-of-way when passing other workers.
- Transfer the tool handle first when passing the tool to others.

Figure 7.4 When working with hand tools, firefighters should work 10 feet (3 m) apart for safety.

Cutting Tools

Cutting tools are primarily used for fireline construction, including cutting brush and small trees **(Figure 7.5)**. Normally, many hours of training and field experience are needed before the required degree of skill in using these tools is attained. If used improperly, these tools can be dangerous; therefore, appropriate caution must be used **at all times**. The most common hand cutting tools for wildland fire fighting are the following:

Figure 7.5 Fireline construction requires the use of cutting tools.

170   Chapter 7 • Tactical Resource Support and Operations

Figure 7.6 A pulaski.

Figure 7.7 A brush hook.

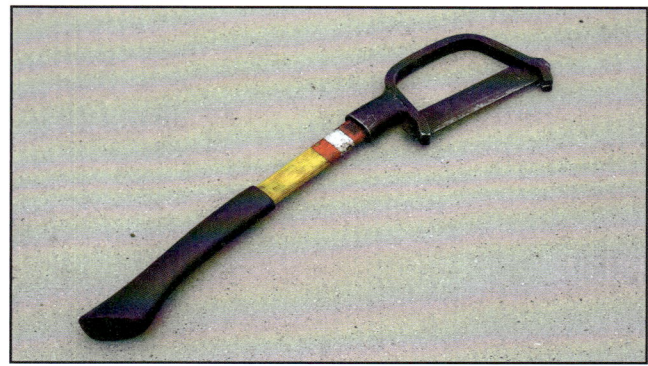

Figure 7.8 A sandvik.

- Axes
- Pulaskis (**Figure 7.6**)
- Brush hooks (**Figure 7.7**)
- Sandviks (**Figure 7.8**)

Scraping Tools

Scraping tools are used for fireline construction and mop-up operations. They can be used to clear away small vegetation and debris to assist in making a fireline. They can also be used to sift through and break up small vegetation and debris. The most common types of scraping tools used in ground cover fire fighting are the following:

- Shovels
- McLeods (**Figures 7.9**)
- Fire rakes-hoes (**Figure 7.10, p. 172**)
- Combination tools (**Figure 7.11, p. 172**)
- Wire brooms

Figure 7.9 A McLeod.

Tactical Resource Support and Operations • Chapter 7   **171**

Figure 7.10 Fire-rake hoes are being used in this photo.

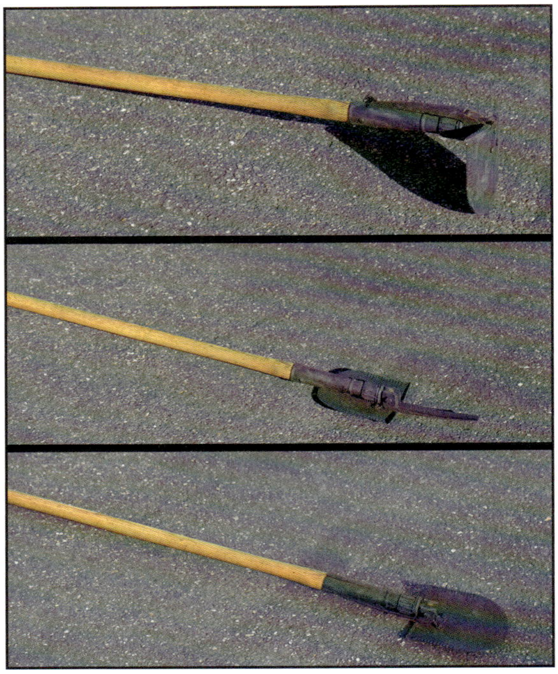

Figure 7.11 A combination tool.

Fire Swatters (Flails)

Fire swatters, sometimes called *fire flails* or *flappers*, are used to smother fires in light fuels such as pasture grasses, pine-needle litter, and light hardwood litter. Flails are very effective when used in conjunction with a backpack pump or fire rake. Flails are used to knock down the flames. The fire is then mopped up with water from the backpack pump or by scraping with the fire rake. Although such things as wet spruce boughs can be used for this purpose, the tool used most often for smothering wildland fires is the fire swatter.

The fire swatter is a long-handled tool with a rubber or neoprene flap attached to one end **(Figure 7.12)**. The flap is usually square in shape with each side 16

Figure 7.12 A fire swatter.

to 24 inches (400 mm to 600 mm) in length. The flap may be replaced if it becomes damaged by heavy use. In use, the flap is swatted or dragged along the edge of a fire **(Figure 7.13)**. However, if the fire is hit too hard, burning embers may be scattered into the unburned area and spread the fire.

Wire Brooms

The wire broom is another tool designed for use in leaf litter, grass, grain, and moss fires. Some wire brooms resemble push brooms; others resemble ordinary straw brooms with wire bristles **(Figure 7.14)**. The wire broom is especially effective in volcanic areas where light, sparse grasses protrude through a layer of small lava rocks. The grasses are literally swept away to create an effective fireline.

Backpack Pumps

Backpack pumps are included in this section because they are manually operated devices that are used to extinguish fire. A backpack pump is used as a form of portable fire extinguisher that carries plain water or a foam/water solution **(Figure 7.15, p. 174)**. It is used to attack small fires and hot spots and to overhaul areas that are not within reach of hoselines. Backpack pumps discharge water when the wearer operates a sliding piston pump in the nozzle or operating a mechanical arm on the side of the unit **(Figure 7.16, p. 174)**. A stream of water is discharged each time the wearer moves the pump handle out and then back in. Though not as common, manually pressurized backpack pumps that permit pressurization of the tank vessel are used in some areas. The actual tank of the backpack pump may be rigid or collapsible.

Depending on local preference, rigid backpack tanks may be stored either full of water or empty. Collapsible backpacks are usually stored empty. Regardless of the type, units that are stored empty should be checked periodically to make sure that all seals are pliable and not dehydrated. All parts should be checked for dirt or rust that may affect their operation. Moving pistons and suction check valves should be kept in good operating condition using a non-oil-based lubricant recommended by the manufacturer.

Figure 7.13 A firefighter uses a fire swatter on a fire in short stubble. *Courtesy of NIFC.*

Figure 7.14 A typical wire broom used in wildland fire fighting.

Figure 7.15 A backpack pump is used as a form of portable fire extinguisher.

Figure 7.16 Backpack pumps discharge water when the wearer operates a sliding piston pump in the nozzle.

Fire and Aviation Management

The US Forest Service captures the following information: "The US Forest Service uses tools in the air to manage fire on the ground. Planes and helicopters are critical tools in managing wildland fire. Although aircraft are often used to fight wildfires, aircraft alone cannot put them out. Firefighters rely on planes and helicopters to:

- Deliver equipment and supplies.
- Deploy smokejumpers and rappellers to a fire.
- Transport firefighters.
- Provide reconnaissance of new fires, fire locations, and fire behavior.
- Drop fire retardant or water to slow down a fire so firefighters can contain it.
- Ignite prescribed fires."

Helicopter Operations Safety Overview

Air operations at wildland fires are safe when pilots and ground personnel use common sense and follow established procedures. Those working in and around operating aircraft must know the applicable safety procedures and follow them without exception. All Forest Service helicopter operations follow Federal Aviation Administration and interagency standards, policies, and safety procedures.

The Federal Aviation Administration (FAA) and organizational policy regulate helicopter landing zones at established heliports or unimproved temporary landing sites. All medical and military service providers must comply with regulations when landing on or off airport property. Fire and emergency services organizations need to be knowledgeable on the requirements of establishing temporary landing zone sites. Coordination may be needed between several agencies in order to effectively secure a temporary landing site. For example, shutting down a roadway will require traffic and crowd control with the assistance of law enforcement. Additionally, holding a temporary landing site for several minutes after lift-off (in case of an emergency with the aircraft) should be coordinated between all involved agencies.

Preplanning with potential helicopter service providers is essential to ensuring safe air and ground operations. Ground crews need training on proper lighting, hazard identification and mitigation, and communications (**Figure 7.17**). The ground crew will identify a temporary landing zone, but the flight crew will approve it. The temporary landing zone size ultimately depends on the size of the helicopter and day/night operations. A good rule to follow is for the landing zone site to be two times the size of the rotor diameter. The flight crew will have the ultimate say of whether to land. The ground crew should have alternate sites pre-identified if the initial location is not approved.

Figure 7.17 Ground crews should be well trained on safe operations on or around helicopter landing zones.

Communications between ground and flight crews should be prearranged. The goal is for ground and flight crews to be able to communicate without the need for relaying information. Some radio systems can be programmed with radio channels or talk groups to facilitate direct communications. When this type of system is not possible, radio frequencies should be identified and readily available. Hand signals are a last resort means of communicating with the flight crew and may be needed in rare situations. Ground crew members should follow standard FAA hand signals as outlined in FAA AC 91-32B, *Safety In and Around Helicopters*.

Helicopters

As stated by the US Forest Service, helicopters are versatile fire management tools. They can be fitted with tanks or buckets to deliver water and fire retardant to the fireline. Large, heavy-lift helicopters can fill their water tanks with a snorkel that siphons water from lakes, rivers, or other sources. Smaller helicopters carry water in buckets that hold between 100 to 400 gallons (400 L to 1 600 L) of water. Each bucket has a release valve on the bottom controlled by the helicopter crew. When the helicopter is in position, the crew releases the water to extinguish hot spots.

Helicopters also deliver fire fighting crews, called *helitack* crews, to fires for initial attack (**Figure 7.18**). Some helitack crews are trained to rappel from helicopters to fires in remote locations.

Helicopters may be used in a variety of other roles in ground cover fire fighting operations. These other roles include the following:

- Performing reconnaissance of the fire scene (including mapping)
- Transporting injured firefighters or civilians from the scene to a medical facility Conducting aerial ignition of backfires and burnout operations
- Filling remote portable water tanks to support relay pumping or hose lays in remote areas

Figure 7.18
Helicopters deliver helitack crews to fires for initial attack.

Fixed-Wing Aircraft

Fixed-wing aircraft are the most effective direct-attack fire fighting resource available. Fixed-wing aircraft have the versatility to start an Initial Attack well before any other resource can reach the fire if it is in a hard-to-reach area. The two most common uses are for fire-scene reconnaissance and air drops of water, foam, or fire retardant. Planes are highly effective in laying down fire retardant lines over fire fuels, thus slowing the spread of the fire and allowing other resources to strengthen firelines. Less frequently, aircraft are also used to drop smokejumpers or cargo. The major cost associated with these units is their use of fuel and retardant costs, in addition to personnel assigned to the aircraft team—the pilots and maintenance crews.

Smokejumpers

A smokejumper is a specifically trained and certified firefighter who travels to wildland fires by aircraft and parachutes to the fire (**Figure 7.19**). On average, smokejumpers parachute in to approximately 400 wildfires per year throughout the western U.S. and Alaska.

Figure 7.19 A smokejumper. *Courtesy of NIFC.*

Smokejumpers perform three primary roles in wildfire management.

1. Parachute in to suppress wildfires while they are still small during "initial attack," or the first response after a wildfire is detected, to prevent them from becoming large, costly, and dangerous to other firefighters and the public.
2. Provide leadership on the fireline and manage "emerging fires," or those that cannot be easily suppressed during initial attack. During the 48 hours it typically takes to mobilize an Incident Management team, smokejumpers often serve as Incident Commanders and other overhead positions (approximately 250 smokejumpers are qualified to serve in various overhead positions).
3. Perform specific short-duration missions on large wildfires that require aerial delivery of firefighters and emergency responders (USDA Forest Service Fire and Aviation Management).

Single-Engine Air Tankers (SEAT)

To realize the full economic and operational effectiveness of single-engine air tankers (SEATs) and to optimize their self-sufficient capabilities, SEAT flight operations should be established as close to the incident as possible using available airports. Therefore, it is crucial that the user be familiar with the operational limitations of these types of aircraft.

The responsibility for planning the most efficient use of SEATs falls directly on the aviation management of the user agency. SEATs are very versatile and can be used from a wide variety of aviation facilities. The using agency should conduct preplanning efforts that include identifying suitable landing sites and operational areas that will promote effective use of the SEAT. Personnel should develop agreements and operational plans for these sites prior to fire season. Some of the criteria that can be used in choosing these sites can be:

- Facilities located in areas with historically high fire occurrences.

- Locations that allow rapid movement of support equipment.
- Locations that can be easily accessed for providing logistical support.
- Areas that have good communications established.
- Facilities that are not subjected to high public use.
- Flight paths over congested areas are minimized.
- Locations that can expand to meet the incident's needs.
- Locations that will help facilitate any security needs.
- Locations that will accommodate aircraft size, type or performance (PMS 506: *NWCG Interagency Single Engine Airtanker Operations Guide*).

NOTE: The tactical use of air assets is a specialized field. The establishment of an *Air Operations Branch* within the incident organization is recommended, and appropriate personnel should be assigned for the coordination and support of these operations.

Air support assets can be ordered through local, regional or state dispatch centers or coordinated through Emergency Operations Centers that have been stood up during ongoing operations. Each jurisdiction should inquire as to what the standard protocols are for ordering national assets or contract resources.

References

National Park Service. Accessed online. https://www.nps.gov

National Wildfire Coordinating Group (NWCG). Accessed online. https://www.nwcg.gov/

National Wildfire Coordinating Group (NWCG) *Interagency Single Engine Air Tanker Operations Guide*. PMS 506. April 2014. Accessed online. https://www.nwcg.gov/publications/506

USDA Forest Service Fire and Aviation Management Operations Guide, PMS 506, April 2014, PMS 506. pdf. Accessed online. https://www.fs.fed.us/fire/people/smokejumpers/

USDA Forest Service. Accessed online. https://www.fs.fed.us/fire/people/handcrews/about_handcrews.html

Chapter 8

Fire Control Tactics

Table of Contents

Backfiring and Burning Out 181
    Backfiring 182
    Burning Out..183

Firing Devices... 184
    Drip Torch...184
    Fusee...185
    Very Pistol/FireQuick.............................186

    Pneumatic Torch.....................................187
    Propane Torch...187
    Terra Torch 187
    Plastic Sphere Dispenser (PSD) 188
    Helitorch..188

References 189

Fire Control Tactics

Chapter 8

Ground cover fires are often controlled by backfiring or burning out to consume fuel between a control line and the main body of a fire. The advantages and disadvantages of each will be explained in this chapter. *Firing devices* are the many devices that are available to start these fires, and these will be described as well.

S-219 *Firing Operations* (Blended) 2014, NWCG Training Development Program, https://onlinetraining.nwcg.gov/node/178 replaces S-234, *Ignition Operations* and serves as an excellent information resource on this topic.

Backfiring and Burning Out

In ground cover fire operations, there are two ways in which fire is used to fight fire. These two ways are known as *backfiring* and *burning out* (**Figure 8.1**). While both backfiring and burning out are considered "firing operations," and both eliminate unburned fuel between a fire's edge and a control line, they are used under different conditions. However, these terms are often misunderstood and interchanged.

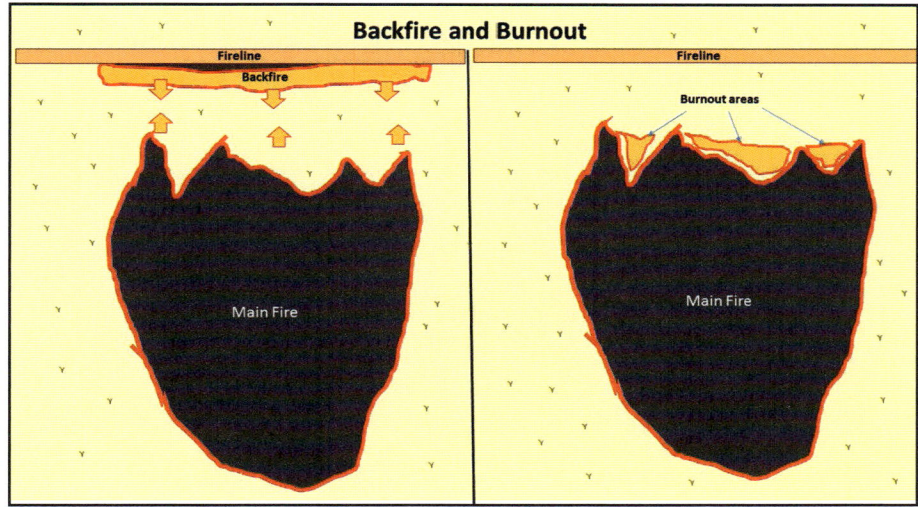

Figure 8.1 Backfiring and burning out are the two ways in which fire is used to fight fire.

Backfiring

Backfiring is associated with indirect attack and is used to stop the spread of a very intense ground cover fire. Backfiring is done from the downwind side of a large fire (the active fire edge) and takes advantage of the natural indraft of the main fire to draw the backfire toward the main fire **(Figure 8.2)**. Although backfiring does widen the defense perimeter, it may also be used to change the force and direction of the convection column.

Figure 8.2 Backfire is used to stop the spread of a very intense ground cover fire. *Courtesy of U.S. Navy Photographer's mate 2nd class Scott Taylor.*

Operations of backfiring include:

- Use only approved trained and experienced firefighters.
- Have the Incident Commander (IC) approve all backfiring operations. The effects of backfiring may influence fire conditions and require modification of the Incident Action Plan (IAP).
- To backfire on *large* incidents, the Operations Section Chief (with IC approval) makes the decision.
- Use fire behavior analysts and incident meteorologists at the division level to place into effect the development of a written plan.

The advantages of backfiring include:

- Reduces fatigue, heat, and smoke exposure to firefighters by letting the backfire do the work of widening the fireline.
- Reduces preparation time with the use of natural barriers/fuel breaks.
- Eliminates fuel ahead of the advancing fire front.
- Reduces fire intensity and spread rate near control lines.
- Alters direction of main fire spread with enough intensity.
- Increases tactical options by allowing fire to do the work of improving fireline.

Judgment and experience must be relied upon to decide when conditions would make a backfire impractical or unsafe. Some of the disadvantages of backfiring include:

- Requires additional preparation, organization, and coordination than does simple line construction.

- Increases acreage burned.
- May influence fire activity in other divisions in a negative way.

Burning Out

A burnout is setting fire inside a control line to widen it or consume fuel between the edge of the fire and the control line **(Figure 8.3)**. It is generally described as an action taken to strengthen and straighten control lines by eliminating fuel between the fire edge and the control line.

The crew boss, a supervisor in charge of usually 16 to 21 firefighters, has the authority to initiate burnout with approval from the division supervisor, Operations Section Chief, or the Incident Commander.

Operations of burning out include:

- An ongoing part of line construction
- Not adversely affecting the actions of others
- Being in conjunction with other divisions and adjoining forces, if necessary
- Keeping adjoining forces informed
- Ensuring that clear communications and intent are understood throughout all levels
- Avoiding to deploy holding forces in unburned fuels ahead of firing operation

Figure 8.3 Burn out operations are an ongoing part of line construction.

> **CAUTION:**
> **Use of fire is dangerous! Maintain communication and coordination at all times. Make sure that you know your agency policy concerning burning out and backfiring.**

The advantages of burning out include the following:

- Strengthens and secures line
- Reduces required holding forces
- Helps minimize amount of line construction required

The disadvantages of burning out include the following:

- May create additional work
- Increases risk of spotting across control lines
- Increases smoke volume and smoke-management issues
- Increases acreage burned

> **NOTE:** Much of the following information is taken or adapted from the following resource: S-219, *Firing Operations* (Blended) 2014, NWCG Training Development Program, https://onlinetraining.nwcg.gov/node/178. (S-219 replaces course S-234, *Ignition Sources*.)

Firing Devices

Devices that are available to start burning out and backfiring are classified as *firing devices*. Firing devices are the tools that you will use to fight fire *with* fire. Firing devices are powerful tools and have the potential to be very dangerous. Operation of any firing device requires a thorough review of the manufacturer's instructions. All personnel must be appropriately trained to use a specific device and wear their fireline personal protective equipment (PPE) with their sleeves rolled down and gloves on.

The following are the firing devices listed in the course S-219 *Module 1 Firing Devices*:

- Drip Torch
- Fusee
- Very Pistol/Fire Quick
- Pneumatic Torch
- Propane
- Terra Torch
- Plastic Sphere Dispenser
- Helitorch

Drip Torch

A drip torch is one of the most common firing devices in use **(Figure 8.4)**. It will work well in almost all fuel types. The fuel used for a drip torch is a mixture typically comprised of a mixture of diesel and gas (3:1 ratio diesel to gas). Personnel must wear full PPE.

To place the drip torch into service:

- Fill the torch with premixed fuel.
- Remove discharge sealing plug and re-screw the plug into the blind-threaded socket on the underside of the lid.
- Position the spout so that wick and loop are pointing away from the handle.

Figure 8.4
Firefighters use a drip torch, which is one of the most common firing devices in use. *Courtesy of U.S. Navy Photographer's mate 2nd class Scott Taylor.*

- Slide lock ring over the spout and screw lock ring on securely.
- Open the breather valve on top of the lid.

  Ignite the drip torch by using:
- Matches/lighter
- Small ground fire
- Another drip torch

  Firing:
- Carry the torch away from your body on the downwind side or upslope side with the wick facing forward.
- Extinguish the drip torch by setting upright. Or, close the breather vale and let the wick burn dry.

> **CAUTION:**
> **Using a gloved hand or blowing out the wick may cause burns to the hands or face.**

Fusee

A fusee is a common firing device and is used for backfiring and burning out **(Figure 8.5)**. Fusees are most effective in light/dry fuels. Fusees burn phosphorous contained within the body of the device. Phosphorous burns very hot (1,400°F [760°C]) and easily ignites grass, twigs, leaves, and other light fuels. A fusee usually burns for 15 to 30 minutes. Firefighters must wear full PPE when working with a fusee.

> **NOTE:** Do not store partially used or broken fusees in your fire pack.

For lighting a fusee, use the following steps:

Step 1: Tape the fusee to a stick or tool handle if it does not have a handle. You will not have to bend over to light the fire.

Step 2: Grip the fusee in one hand, and remove the striker cap by the tapered end.

Step 3: Scrape the striker end sharply against the ignition end of the fusee in a downward motion, away from the body.

Step 4: Holding the fusee downwind, turn your head to the side when striking the fusee.

Step 5: Hold the fusee away from the body with the lighted end down so that burning phosphorous does not drip onto your hand.

Step 6: Stay in the black once the backfire is started.

Step 7: Extinguish the fusee by tapping the burning end on a noncombustible surface or turning upright until it burns out.

Figure 8.5 Fusees are most effective in light/dry fuels.

Step 8: Dispose of fusees properly. Do not leave them where livestock or other animals can eat them.

> **CAUTION:**
> **If not handled properly when ignited, fusees (comprised of phosphorous) are subject to spattering and can cause severe burns. The smoke can also cause headaches.**

> **NOTE:** As a last resort, a fusee can be used to burn out an emergency safety zone in light fuels. When doing so, this will not compromise the safety of others.

Very Pistol/FireQuick

These are specialized pistols adapted to shoot flares up to 375 feet (120 m) (**Figure 8.6**). Flare will burn approximately 8 seconds in dry, light and continuous ground fuels.

Figure 8.6 A Very Pistol. *Courtesy of the National Interagency Fire Center.*

Firing:

- Initiate firing from positions that allow clear shots.
- Notify all personnel in the immediate area prior to firing.
- Pick up and dispose of all spent cartridges.

FireQuick is another specialized pistol developed to shoot either HotShot or Stubby Flares. A supervisor ensures that the pistol and cartridges are always under the control of qualified personnel at all times.

The two most common loads fired by the FireQuick:

- #6, Purple-Violet Box. For most uses when launching HotShot flares. Winchester Industrial Loads, 22 caliber.
- #7 Gray Box. When launching Stubby flares or when launching a HotShot flare for longer distances. These loads can damage the fired end of the flare and cause more wear and tear on the launcher than #6 loads. Winchester Industrial Loads, 22 caliber.

Tips:
- Use the cylinder retaining pin to dislodge fired cartridges caught in the cylinder.
- To increase dexterity, use NomeX® flight gloves to handle, load, and fire the pistol.
- Always follow your agency's policy regarding PPE.

> **CAUTION:**
> Always point the launcher in a safe direction away from personnel and equipment.

Pneumatic Torch

The pneumatic torch assembly is comprised of the following:
- Torch assembly
- Fuel tank
- A compressed air cylinder that can be ATV mounted or carried as a backpack.

The fuel is a mixture of diesel and gas. Depending on the model of the pneumatic torch, it can dispense fuel 8 to 20 feet (2.5 m to 6 m). This torch can be effective in all fuels.

> **NOTE:** Monitor the pneumatic torch due to the capacity of putting too much heat on the control line.

> **CAUTION:**
> Inform operator of the radiant heat exposure.

Propane Torch

Typically, a propane torch is comprised of the following:
- Torch assembly
- Propane cylinder
- Hose
- Regulator or flame adjustment valve

This device can be handled or can be mounted in the back of pickup truck or trailer. Most torches use propane vapor as the fuel, although some are designed to use liquid propane. This torch can be effective in all fuels.

> **CAUTIONS:**
> **Propane torch:**
> - Exposes the operator to radiant heat.
> - Has the potential to put too much heat on the control line.

Terra Torch

This apparatus produces a high-volume flame with a range of approximately 100 feet (30 m), depending on pump pressure and terrain. The Power Flame Thrower/Terra Torch is comprised of the following:

- Fuel tank
- Torch assembly
- Pump
- Electric-propane igniter
- Wand

This device is effective in all fuels, especially brush and moist fuels, due to its longer residence time. The Terra Torch may be installed in the back of a pickup bed, trailer, boat or heavy equipment (**Figure 8.7**). It can be very effective when firing off roads and dozer lines.

Figure 8.7 The Terra Torch can be very effective when firing off roads. *Courtesy of Candice Stevenson/ US Fish and Wildlife Service (USFWS).*

CAUTIONS:
The Terra Torch:
- **Exposes the operator to radiant heat**
- **Produces hazardous fumes**
- **Needs to be monitored due to the capacity of putting too much heat on the control line**

Plastic Sphere Dispenser (PSD)

The PSD is comprised of a glycol (antifreeze) injection unit that is mounted internally in a helicopter with an external chute. Operation of the PSD requires a qualified individual Plastic Sphere Dispenser Operator (PSDO). This system eliminates the handling of raw materials, external loads, and reduces the number of ground forces needed. The PSD Dispenser has a lower operating cost than a Helitorch and is most effective in dry, light, continuous fuels and aquatic areas like marshes or swamps. Can be used to ignite a large area in a short amount of time.

Helitorch

A Helitorch is comprised of the following:
- Torch assembly

- Propane
- Tank
- Pumps
- Valves
- Wiring harness
- Ignition tip. This device is carried as an external load on a helicopter.

    Helitorch Capabilities:

- Ignites fuel with higher fuel temperatures but will not burn wet fuels.
- Expands your prescription window; however, it is also prone to equipment failure and typically requires a back-up firing device.
- Can be used to ignite a large area in a short amount of time and is capable of burning standing brush and fuel types with a little or no ground fuels **(Figure 8.8)**.

Figure 8.8 A helitorch is carried as an external load on a helicopter. *Courtesy of InciWeb-Incident Information System. NWCG. Cliff Creek Fire.*

- Requires specialized personnel to operate.

For more information and procedures on plastic sphere dispensers, Helitorchs, or other firing devices, visit: http://www.nwcg.gov/pms/pubs/443/pms443.pdf

References

S-219 Module1 - NWCG

https://training.nwcg.gov/courses/s219/OCM01/index.html

Glossary

Glossary

A

Anchor Point — Point from which a fireline is begun; usually a natural or man-made barrier that will prevent fire spread and the possibility of a crew being "flanked" while constructing the fireline. Typical anchor points are roads, lakes, ponds, streams, earlier burns, rock slides, and cliffs.

B

Backfire — Fire set along the inner edge of a control line to consume the fuel in the path of a wildland fire and/or change the direction of force of the fire's convection column (NWCG).

Backfiring — A tactic associated with indirect attack, intentionally setting fire to fuels inside the control line to slow, knock down, or contain a rapidly spreading fire. Backfiring provides a wide defense perimeter and may be further employed to change the force of the convection column. Backfiring makes possible a strategy of locating control lines at places where the fire can be fought on the firefighter's terms. Except for rare circumstance meeting specified criteria, backfiring is executed on a command decision made through line channels of authority (NWCG).

Black — Area already burned by a wildland fire.

Booster Line — Noncollapsible rubber-covered, rubber-lined hose usually wound on a reel and mounted somewhere on an engine or water tender. This hose is most commonly found in ½-, ¾-, and 1-inch (13 mm, 19 mm, and 25 mm) diameters and is used for extinguishing low-intensity fires and mop-up. *Also called* Hard Line.

Burning Out — Setting fire inside a control line to consume fuel located between the edge of the fire and the control line (NWCG).

C

Command Staff — The command staff consists of the Information Officer, Safety Officer and Liaison Officer. They report directly to the Incident Commander and may have an assistant or assistants, as needed (NWCG).

Conduction — Heat transfer through a material from a region of higher temperature to a region of lower temperature (NWCG).

Control Line — Inclusive term for all constructed or natural barriers and treated fire edges used to control a fire (NWCG).

Convection — Transfer of heat in an upward vertical motion which can dry and ignite fuels above and adjacent to the fire. The most common visual indicator of convection is the smoke column, which consists of hot air, gases, embers, and debris.

Creeping — Fire burning with a low flame and spreading slowly (NWCG).

Crown fire — Fire that advances from top to top of trees or shrubs driven by a number of factors, intensity of the surface fire, slope, spacing of the crowns, and wind.

D

Direct Attack — Any treatment applied directly to burning fuel such as wetting, smothering, or chemically quenching the fire or by physically separating the burning from unburned fuel (NWCG).

E

Escape Route — A preplanned and understood route firefighters take to move to a safety zone or other low-risk area. When escape routes deviate from a defined physical path, they should be clearly marked (flagged) [NWCG].

Evacuation — An organized, phased, and supervised withdrawal, dispersal, or removal of civilians from dangerous or potentially dangerous areas, and their reception and care in safe areas (NWCG).

F

Fingers — Long, narrow extensions of a fire projecting from the main body.

Fire Edge — The boundary of the burned or burning material at any given time.

Fireline — Part of a control line that is scraped or dug to mineral soil; also, a general term for the area where fire fighting activities are taking place, the wildland equivalent of the term "fireground" as used in structural fire fighting.

Fire Shelter — An aluminized cloth tent that offers protection in a fire entrapment situation by reflecting radiant heat and providing a volume of breathable air (NWCG).

Fire Whirl — Spinning vortex column of ascending hot air and gases rising from a fire and carrying aloft smoke, debris, and flame. Fire whirls range in size from less than one foot to over 500 feet in diameter. Large fire whirls have the intensity of a small tornado (NWCG).

Flame Height — The average maximum vertical extension of flames at the leading edge of the fire front. Occasional flashes that rise above the general level of flames are not considered. This distance is less than the flame length if flames are tilted due to wind or slope (NWCG).

Flame Length — The distance between the flame tip and the midpoint of the flame depth at the base of the flame (generally the ground surface), an indicator of fire intensity (NWCG).

Flank Attack — Attacking a fire by working along the flanks either simultaneously or successively from an anchor point.

Flanking Fire Suppression — Attacking a fire by working along the flanks either simultaneously or successively from a less active or anchor point and endeavoring to connect two lines at the head (NWCG).

Flanks of a Fire — Parts of a fire's perimeter that are roughly parallel to the main direction of spread (NWCG).

G

Green — Area of unburned fuels, not necessarily green in color, adjacent to but not involved in a wildland fire.

H

Hand Crew — A number of individuals who have been organized and trained and are supervised principally for operational assignments on an incident (NWCG).

Hard Line — *See* Booster Line.

Head of a Fire — The most rapidly spreading portion of a fire's perimeter, usually to the leeward or up slope (NWCG).

Heel — Rear portion of a wildland fire. *Also called* Rear.

Hose Lay — (1) Arrangement of connected lengths of fire hose and accessories on the ground at a wildland fire beginning at the first pumping unit and ending at the point of water delivery. (2) Connected lengths of hose from water source to pumping engine.

Hotshot Crew — Intensively trained fire crew used primarily in hand line construction (Type-1) [NWCG].

Hot-Spotting — Checking the spread of fire at points of more rapid spread or special threat. Is usually the initial step in prompt control, with emphasis on first priorities (NWCG).

I

Incident Action Plan (IAP) — Contains objectives reflecting the overall incident strategy and specific tactical actions for the next operational period; may be oral or written. When written, the plan may have a number of forms as attachments.

Incident Commander — Person in charge of and responsible for the management of all incident operations.

Incident Command System (ICS) — A standardized on-scene emergency management concept specifically designed to allow its user(s) to adopt an integrated organizational structure equal to the complexity and demands of single or multiple incidents, without being hindered by jurisdictional boundaries.

Indirect Attack — Controlling the fire by locating the control line along natural firebreaks some distance from the approaching fire and burning out the intervening fuels.

Inversion — Atmospheric inversion. The departure from the usual increase or decrease with altitude of the value of an atmospheric property. In fire management usage, nearly always refers to an increase in temperature with increasing height. Also, the layer through which this departure occurs (also called inversion layer.) The lowest altitude at which the departure is found is called the base of the inversion (NWCG).

Island — Unburned area within a fire perimeter.

L

Lookout — (1) Person designated to detect and report fires from a vantage point. (2) Location from which fires can be detected and reported. (3) Fire crew member assigned to observe the fire and warn the crew when there is danger of becoming trapped.

M

Mobile Attack — Suppressing fire along a fire edge by driving mobile apparatus along the perimeter and simultaneously applying fire streams to knock down the fire. *Also called* Pump and Roll.

O

Operations Section Chief — This ICS position is responsible for supervising the Operations Section. Reports to the Incident Commander and is a member of the General Staff. This position may have one or more deputies assigned.

Origin — Point of original ignition of a fire.

P

Parallel Attack — Constructing a fireline parallel to a wildland fire's edge. After the line is constructed, the fuel inside the line is burned out.

Patrol — (1) To travel over a given route to prevent, detect, and suppress fires. (2) To go back and forth vigilantly over a length of control line during and/or after construction to prevent slopovers, suppress spot fires, and extinguish overlooked hot spots.

Perimeter — Entire outer edge or boundary of a fire.

Pincer Attack — Direct attack around a fire in opposite directions by two or more attack units with the ultimate intent of pinching off (stopping) the head of the fire.

R

Radiation — Transfer of heat in straight lines through a gas or vacuum other than by heating of the intervening space.

Rate of Spread — The relative activity of a fire in extending its horizontal dimensions. It is expressed as rate of increase of the total perimeter of the fire, as rate of forward spread of the fire front, or as rate of increase in area, depending on the intended use of the information. Usually it is expressed in chains or acres per hour for a specific period in the fire's history.

Running Fire — Behavior of a fire spreading rapidly with a well-defined head, usually associated with influences of wind and/or slope.

S

Safety Zone — An area cleared of flammable materials used for escape in the event the line is outflanked or in case a spot fire causes fuels outside the control line to render the line unsafe. In firing operations, crews progress so as to maintain a safety zone close at hand allowing the fuels inside the control line to be consumed before going ahead. Safety zones may also be constructed as integral parts of fuel breaks; they are greatly enlarged areas which can be used with relative safety by firefighters and their equipment in the event of blowup in the vicinity (NWCG).

Shelter in Place — Remaining in a structure or vehicle when a fire moves through rather than attempting to use roads that may be blocked or untenable because of fire; opposite of evacuation.

Size-Up — Ongoing process of observation and evaluation of existing factors that are used to develop objectives, strategy, and tactics for fire suppression.

Slopover — A fire edge that crosses a control line or natural barrier intended to confine the fire (NWCG).

Smoldering — Fire burning without visible flame and barely spreading.

Spot Fire — Fires starting outside the perimeter of a main fire typically caused by flying sparks or embers. *Also see* Spotting.

Spotting — Behavior of a fire producing sparks or embers that are carried by the wind and which start new fires beyond the zone of direct ignition by the main fire (NWCG).

Structural Triage — Process of inspecting and classifying structures according to their *defensibility/indefensibility* based on their situation, their construction, and the immediately adjacent fuels.

T

Torching — The burning of the foliage of a single tree or a small group of trees, from the bottom up (NWCG).

Turn Down — A situation where an individual has determined that he or she cannot undertake an assignment as given and is unable to negotiate an alternative solution.

U

Unified Command — Unified team effort in the Incident Command System that allows all agencies with responsibility for the incident, either geographical or functional, to manage the incident by establishing a common set of incident objectives and strategies. This is accomplished without losing or abdicating authority, responsibility, or accountability. In unified command there is a single incident command post and a single operations chief at any given time.

Reference

National Wildfire Coordinating Group (NWCG) *Glossary of Wildland Fire Terminology*. Accessed online. https://www.nwcg.gov/glossary/a-z

Index

Index

A

Access, ground cover/urban interface operations, 121–122
Accountability, ground cover/urban interface operations, 119–120
Acronym, SMART, 77, 85
Administration, Finance/Administration Section, 67
Aerial fuels, 5
Agency policy, personal protective equipment maintenance and use, 40
Aggressive fire fighting, safety during, 26
AHJ. *See* Authority having jurisdiction (AHJ)
Aid agreements, 73–74
Air Operations Branch, 178
Air temperature, 8
Alertness during life-threatening situations, 25
Anchor point of firelines
    dangers of not using, 29
    defined, 91, 97
    examples, 87
    for firefighter safety, 167
    fireline construction, 92–96
    fireline width, 91–92
Animals
    animal trap hazards, 30
    care of, during operations, 134
    hazards during fire operations, 30
Apparatus. *See* Fire apparatus; Heavy equipment
Aspect of topography, 12–13
Assignments (ICS 204 Form), 77
Atmospheric stability
    defined, 9
    fire whirls, 11
    inversion, 10
    nighttime (radiation), 10
    stable atmosphere, 10
    subsidence inversion, 10–11
    unstable atmosphere, 11
Authority, delegation of, 68–69
Authority having jurisdiction (AHJ)
    delegation of authority, 69
    Incident Management System, 59
Automatic aid agreement, 73
Aviation hazard (SAFECOM) form, 27
Aviation management, 174–178
    fixed-wing aircraft, 176
    helicopter operations safety, 174–175
    helicopters, 175–176
    single-engine air tanker (SEAT), 177–178
    smokejumpers, 176–177

B

Backfire, 114, 149
Backfiring
    control lines for firing operations, 144–145
    defined, 143, 149
    tactics, 181–183
    timing and coordination, 144
    uses for, 143–144
Backpack pump, 173–174

Barriers to fire spread
    man-made barriers, 14
    natural barriers
        control line, 115
        fire spread, 14
        as safety zone, 24, 33
Behavior of fire. *See* Fire behavior
Black (burn) area, 83, 97, 110
Booster (hard) line, 111, 116–117, 150
Boots, 36
Box canyon, 14
Branch, NIMS-ICS organizational functions, 60–61
Bridges, apparatus safety on, 46
Briefings
    field-level, 76
    incident briefing (ICS 201 Form), 77
    operational period, 76
    purpose of, 75
    Section-level, 76
    staff-level, 76
    topics, 75–76
Brush fires, 3
Brush hooks, 49, 171
Building. *See* Structure triage
Burn (black) area, 83, 97, 110
Burned area as safety zone, 24
Burning out
    advantages/disadvantages of use, 183
    defined, 150, 183
    operations, 183
    purpose of, 181
    uses for, 143–144

C

Canyons, 14
Carbon monoxide, 30
Cavitation, 157
Center for Public Safety Excellence, 73
Chain of command, NIMS-ICS, 59, 78
Chain saws used for fireline construction, 95–96
Chimney effect, 12, 14
Chutes, 12
Clandestine drug labs, 51
Class A foam
    dry foam, 141
    fluid foam, 141
    injecting air into, 140
    pretreatment of structures, 134, 141–142
    properties, 141
    wet foam, 141
    wet lines, 144–145
Combination deployment of resources, 86
Combination tools, 171, 172
Command, ground cover/urban interface operations, 119–120
Command Section, NIMS-ICS, 61–62
Command Staff, NIMS-ICS organizational functions, 60, 62–64, 78
Common Denominators of Fire Behavior on Tragedy Fires, 43–44

Communication
- engine signals, 50
- hand signals, 50
- lack of, as danger to firefighters, 28
- lookouts and crew leaders, 32
- lookouts, communications, escape routes, and safety zones (LCES), 31–35
- MAYDAY, 75
- NIMS-ICS, 74–76
  - briefings, 75–76
  - formal communications, 75
  - informal communications, 75
- radios, 32, 49–50
- *Standard Firefighting Orders*, 25

Compactness of fuels, 5
Compensation/Claims Unit, 67
Constructed safety zone, 33
Containment of fires around structures, 138
Continuity of fuels, 5
Control, ground cover/urban interface operations, 119–120
Control line
- defined, 97
- firing operations, 144–145
- indirect attack, 115
- location of, 87
- natural barriers, 115

Convection and fire behavior, 13
Cost Unit, 67
Crews. *See also* Teams
- danger of being uninformed, 28
- defined, 72
- fire crews, 165–167
- hand crew, 113, 150, 166–167
- helitack crew, 176
- hotshot, 113, 150
- Interagency Hotshot Crew (IHC), 166
- *Standard Firefighting Orders* for control of, 26

Crown fires, 4, 5
Cutting tools, 170–171

D

Daylight view of areas, 28
Deaths, line-of-duty, 31
Debris hazards during fire operations, 30
Decisive actions during life-threatening situations, 25
Defensive deployment of resources, 86
Delegation of authority, 68–69
Demobilization Unit, 65
Deputy Incident Commander, 63
Direct attack, 86, 88–89, 97
Division, NIMS-ICS organizational functions, 61
Division of labor, 70
Division Supervisor, 120
Documentation Unit, 65
Downslope winds, 7
Dozers
- fireline construction, 92, 96, 97
- use in ground cover fire fighting, 48, 168, 169

Drainages, impact on fire behavior, 12
Drip torch, 184–185
Drug labs, 51
Dry foam, 141
Dry hydrant, 157
Dumping capabilities of water tenders/ground tankers, 160–161

E

Egress, ground cover/urban interface operations, 120–122, 126
18 Watchout Situations, 27–30
Electrical power lines, 30, 53
Electrified fences, 30
Engines. *See also* Fire apparatus
- backing in for safety, 126, 127
- cab as safety zone, 33–35
- engine signals for communication, 50
- Type 3 fire engine, 107
- Type 4 fire engine, 107–108
- Type 5 fire engine, 108
- Type 6 fire engine, 108
- Type 7 fire engine, 108
- types of wildland engines, 107–108

Equipment. *See* Heavy equipment
Escape route
- cautions, 25
- defined, 32, 55
- identification for safety, 28, 32
- NWCG Flagging Standards, 32
- *Standard Firefighting Orders*, 24

Evacuation, 121, 150
Explosives
- fire operation hazards, 30
- fireline construction hazards, 132

Extinguishment, hit and run tactics, 143

F

FAA (Federal Aviation Administration), 175
FAA AC 91-32B, *Safety In and Around Helicopters*, 175
FAE (Fire Apparatus Engineer), 158
Falling debris hazards during fire operations, 30
Federal Aviation Administration (FAA), 175
Fences, electrified, 30
Field-level briefings, 76
Finance/Administration Section–Finance/Administration Section Chief, 67
Fingers of a fire, 82, 98
Fire, parts of, 81–83
Fire apparatus. *See also* Heavy equipment
- backing engines in for safety, 126, 127
- confronting the fire at the structure, 137–138
- engine cab as safety zone, 33–35
- engine signals for communication, 50
- extinguishment and follow up, 143
- fire control tactics, 108–115
  - flank attack, 109
  - frontal attack, 110
  - hotspotting, 113, 150
  - indirect attack, 114–115
  - mobile attack, 105, 110–111, 150
  - pincer attack, 109, 150
  - tandem attack, 112–113
- fireline construction, 130–132
- firing operations, 143–145
- foam use, 140–142
- follow-up, 146–147
- ground cover/urban interface operations, 118–127
  - command, control, and accountability, 119–120
  - ingress and egress, 120–122
  - structure triage, 122–127
- hit and run tactics, 142–143
- hose lays, 115–118

hoselines, 135–136
NFPA® 1906 standard, 105
nozzles, 137
pets and livestock, 134
pretreatment of structures, 134
private vehicles, 133
public relations, 147–149
resource typing (NIMS-ICS), 105–108
retreating and returning, 142–143
safety guidelines, 44–47
    capabilities and limitations, 45
    general guidelines, 44
    hoseline safety guidelines, 45
    off-road guidelines, 44–46
    personnel transport, 46–47
structural apparatus for ground cover fires, 103–105
structure and site preparation, 128–130
structure protection tactics, 134–135
types of wildland engines, 107–108
water use, 139–140
Fire Apparatus Engineer (FAE), 158
Fire behavior, 3–17
    air temperature, 8
    atmospheric stability, 9–11
    characteristics, 15–16
    Common Denominators of Fire Behavior on Tragedy Fires, 43–44
    crown fires, 4, 5
    firelines, 14–15
    fire whirl, 11, 16
    flame length, 4, 15
    ground cover fuels, 5–6
    ground fires, 4, 5
    impact on structure triage, 124, 125
    ladder fuels, 5
    local terrain features, 12, 28
    precipitation, 9
    radiation, 10, 17
    rate of spread, 3, 17
    relative humidity, 8–9
    situational awareness concerning, 41
    spotting, 14, 17
    Standard Firefighting Orders, 24
    surface fires, 4, 5
    topography, 11–14
        aspect, 12–13
        defined, 11
        slope, 13
        steepness, 12
        terrain, 13–14
    torching, 15, 17
    weather factors, 6, 9
    wind factors, 6–8
        general winds, 7
        impact on fires, 7
        local winds, 7–8
Fire control strategies, 86–97
    anchor point, 91–97
        defined, 91, 97
        examples, 87
        fireline construction, 92–96
        fireline width, 91–92
    changing the plan, 87
    direct attack, 86, 88–89, 97
    firelines, 90–91
    flank attack, 87
    indirect attack, 86, 89–90, 98
    initiating operations, 87
    line construction with mechanized equipment, 96–97
    parallel attack, 87, 98
Fire control tactics, 108–115
    flank attack, 109
    frontal attack, 110
    hotspotting, 113, 150
    indirect attack, 114–115
    mobile attack, 110–111, 150
    pincer attack, 109, 150
    tandem attack, 112–113
Fire crews, 165–167
Fire edge, 82, 98
Fire extinguisher, backpack pump, 173–174
Fire flail, 172–173
Fire Gel, 134
Fire hydrants, 155–157
Fire shelter
    defined, 37, 55
    guidelines for use, 37–38
    inner layer, 37
    location for deployment, 39–40
    mechanics of, 38–39
    outer layer, 37
Fire stop, 14–15
Fire swatter, 172–173
Fire triangle, 5
Fire whirl, 11, 16
Firefighter safety and survival, 21–55
    Common Denominators of Fire Behavior on Tragedy Fires, 43–44
    18 Watchout Situations, 27–30
    electrical power lines, 30, 53
    fireline construction anchor point, 167
    hazardous materials situations, 51–52
    hazards during fire operations, 30
    heavy equipment operations, 47–51
    impact on structure triage, 126
    lightning hazards, 30–31, 42–43
    lookouts, communications, escape routes, and safety zones (LCES), 31–35
    operational leadership, 21–22
    personal protective equipment. *See* Personal protective equipment (PPE)
    responsibility for, 21
    risk management, 22–26
    rules of engagement, 40–43
        lightning, avoiding, 42–43
        look around, 42–43
        look down, 42
        look up, 41–42
        situational awareness, 40–41
        30/30 rule, 43
    snag safety, 54
    Standard Firefighting Orders, 23–26
    training and preparation, 165
    turn down (refusal of risk), 26–27, 55
    wildland fire apparatus safety, 44–47
Firelines
    anchor point, 29, 167
    chain saws for construction, 95–96
    construction, 92–96, 130–132
        anchor point, 167
        breaks in fuel, 131
        chain saws, 95–96

effectiveness of fireline, 15
flammable and explosive hazards, 132
intermediate fuels, 131
leapfrogging, 93
with mechanized equipment, 96–97
progressive line construction (one-lick method), 93–94
rate of, factors, 96
typical tool order, 94–95
yard accumulation, 131–132
dangers of sleeping near, 30
defined, 14, 55
downhill, hazards of building, 29
locating and developing, 90–91, 167
spot fires, 29
width of, 91–92
FireQuick, 186–187
Fire-rake hoe, 171, 172
Firing devices, 184–189
drip torch, 184–185
fusee, 185–186
Helitorch, 188–189
Plastic Sphere Dispenser (PSD), 188
pneumatic torch, 187
propane torch, 187
Terra Torch, 187–188
Very Pistol/FireQuick, 186–187
Firing operations, 143–145
backfiring, 143–145, 149
burning out, 143–144, 150
firing and holding, 145
S-219 *Firing Operations*, 181
Fixed-wing aircraft, 176
Flagging Standards, 32
Flail, 172–173
Flame length of fire
defined, 16
fire behavior, 4, 15
impact on structure triage, 125
Flammable hazards and fireline construction, 132
Flank attack, 87, 109
Flanking fire suppression, 109, 150
Flanking the fire, 109
Flanks of a fire, 82, 98
Flapper, 172–173
Fluid foam, 141
Foam. *See* Class A foam
Follow-up
hit and run tactics, 143
patrol, 147, 150
structure protection, 146–147
Forest fires, 3
Formal communications, 75
Frontal attack/assault, 29, 110
Frontal inversion, 10
Fuel for fires
burning characteristics, 5–6
characteristics for situational awareness, 41
fireline construction and, 131–132
ground cover, 5–6
impact on structure triage, 124–125
ladder fuels, 5
removing and trimming around structures, 130
rolling material on hillside, 29
structures, 129
unburned fuel between you and the fire, 29

wood piles, 131
wooden fences, 131
yard accumulation, 131–132
yard furniture, 131
Fusee, 185–186

G
Gas, turning off at the source, 133
General Staff, NIMS-ICS organizational functions, 60
General winds, 7
Gloves, 37
Grass fires, 3
Green area, 83, 98
Green zone, 83
Ground cover fuels, 5–6
Ground cover/urban interface operations
fire apparatus, 118–127
command, control, and accountability, 119–120
ingress and egress, 120–122
structure triage, 122–127
parts of a fire, 81–83
structural apparatus used on, 103–105
Ground fires, 4, 5
Ground hazard (SAFENET) form, 27
Ground sweep nozzles, 162
Ground tanker, 160–162
Groups
Group Supervisor, 120
NIMS-ICS organizational functions, 61
NWCG. *See* National Wildfire Coordinating Group (NWCG)
work groups, 70

H
Hand crew, 113, 150, 166–167
Hand signals, 50
Hand tools
backpack pumps, 173–174
carrying guidelines, 170
cutting tools, 170–171
for fighting ground cover fires, 170
fire swatters (flails), 172–173
safety, 49
scraping tools, 171–172
wire brooms, 173
Hard (booster) lines, 111, 116–117, 150
Hard hat, 37
Hazardous materials
clandestine drug labs, 51
fire operation hazards, 30
firefighter safety tips, 51–52
hazards during fire operations, 30
operation guidelines, 52
scene management, 52
Head of a fire, 82, 98
Heart failure during ground cover fires, 31
Heat buildup, reducing, 140
Heat traps, impact on triage, 123
Heat wave, 140
Heavy equipment, 47–51. *See also* Fire Apparatus
contract for use of, 168
dozers
fireline construction, 92, 96, 97
use in ground cover fire fighting, 48, 168, 169

firefighter safety tips, 50
fireline construction, 96–97
hand signals during use of, 50
hand tool safety, 49
hazards during fire operations, 48
operations safety, 48–51
radio communication, 49–50
road graders (maintainers), 48, 169
safety procedures, 96–97
seeing and being seen, 48
tractor-plows, 48, 168–169
types of equipment, 48
underground utilities, 50–51
uses for, 167
vehicle requirements, 47–48
Heel of a fire, 82, 98
Helicopters, 174–176
Helitack crew, 176
Helitorch, 188–189
Hillsides, apparatus safety on, 45
Hit and run tactics, 142–143
Holding firing operations, 145
Hoselines
    hard (booster) lines, 111, 116–117, 150
    hose lays, 115–118
        booster (hard) line, 111, 116–117
        components, 115–116
        defined, 150
        progressive hose lays, 103–104, 117–118
    safety guidelines, 45
    working for structural protection, 135–136
Hotshot crew, 113, 150
Hotspotting, 113, 150
Humidity. *See* Relative humidity
Hydrants, 155–157

I
IAP. *See* Incident Action Plan (IAP)
IC. *See* Incident Commander (IC)
ICP (Incident Command Post), 64
ICS. *See* Incident Command System (ICS)
IHC (Interagency Hotshot Crew), 166
Incident Action Plan (IAP)
    assigned incident tasks, 77
    defined, 76, 78
    development of, 85–86
    elements, 77
    Incident Commander roles, 62
    incident priorities, 85
    NIMS IAP planning process, 77
    resource deployment, 86
    verbal, 77
Incident briefing, 77
Incident Command Post (ICP), 64
Incident Command System (ICS). *See also* National Incident Management System-Incident Command System (NIMS-ICS)
    defined, 78
    ICS 201 Form (incident briefing), 77
    ICS 202 Form (incident objectives), 77
    ICS 203 Form (organization), 77
    ICS 204 Form (assignments), 77
    ICS-100 training, 60
    ICS-200 training, 60
    ICS-300 training, 60
    ICS-400 training, 60
    purpose of, 59–60
Incident Commander (IC), 62–64
    defined, 62, 78
    Deputy Incident Commander, 63
    Incident Command Post, 64
    Incident Safety Officer, 63–64
    Liaison Officer, 64
    Public Information Officer, 64
    size-up, 84–85
    turn down (refusal of risk), 27
Incident management, 68–77
    delegation of authority, 68–69
    incident action plan, 76–77
    incident communications, 74–76
    management by objectives, 71–72
    organizational flow chart, 69
    organizational principles, 69–70
    resource management, 72–74
Incident objectives (ICS 202 Form), 77
Incident Response Pocket Guide (IRPG), 21, 50
Incident Safety Officer (ISO), 63–64
Indirect attack
    backfiring, 182
    dangers during, 89–90
    defined, 98
    tactics, 114–115
    uses for, 86, 89
Informal communications, 75
Ingress, ground cover/urban interface operations, 120–122, 126
Insect hazards, 30
Instructions
    clarity of, for safety, 28
    Standard Firefighting Orders, 26
Insurance Services Office (ISO), 73
Interagency Hotshot Crew (IHC), 166
Interagency Hotshot Operations Guide, 166
Interagency Single Engine Airtanker Operations Guide, 178
Inversion
    defined, 10
    frontal, 10
    marine, 10
    nighttime (radiation), 10
    subsidence inversion, 10–11
    temperature and relative humidity, 9
IRPG (Incident Response Pocket Guide), 21, 50
Islands, 82, 98
ISO (Incident Safety Officer), 63–64
ISO (Insurance Services Office), 73

L
Ladder fuels, 5
Ladders as on-site resource, 129
Lakes as water source, 157
Law enforcement
    evacuation authority, 121
    traffic control, 121–122
Leadership
    operational, 21–22
    supervision, 22, 120
Leapfrogging, 93
Liaison Officer, 64
Lighting for structural operations, 133

Lightning
   avoiding, 42–43
   precautions for protection from, 30–31
   30/30 rule, 43
Line-of-duty deaths during ground cover fires, 31
Livestock, during ground cover/urban interface operations, 134
Local winds, 7–8
Logistics Section–Logistics Section Chief, 66
Look Around
during heavy equipment operations, 51
rules of engagement, 42–43
Look Down
   during heavy equipment operations, 51
   rules of engagement, 42
Look Up
   during heavy equipment operations, 51
   rules of engagement, 41–42
Lookouts
   defined, 31, 55
   function, 29, 32
   lookouts, communications, escape routes, and safety zones (LCES), 31–35
   purpose of, 31, 55
   Standard Firefighting Orders, 25

M

Maintainers (road graders), 48, 169
Management by objectives, 71–72
Man-made barriers to fire spread, 14
Marine inversion, 10
MAYDAY, 75
McLeod, 49, 171
Media, public relations duties, 148–149
Memorandum of Understanding (MOU), 68, 78
Mobile attack, 105, 110–111, 150
Mobile water operations, 158–162
   apparatus, 158
   ground sweep nozzles, 162
   nurse tender operations, 160
   portable water tanks, 159
   pumping and dumping, 160–161
   remote control nozzles, 162
   safe operation, driving, off-road use, 158
   tender/tanker operations, 161–162
   water shuttle operations, 159–160
Moisture content of fuels, 6
MOU (Memorandum of Understanding), 68, 78
Mutual aid agreement, 73

N

Narrow canyons, 14
National Fire Protection Association®, aid agreements, 73. *See also specific NFPA® standard*
National Incident Management System–Incident Command System (NIMS-ICS), 59–78
   chain of command, 59, 78
   Command Section, 61–62
   Finance/Administration Section–Finance/Administration Section Chief, 67
   Incident Commander (IC), 62–64
      defined, 78
      Deputy Incident Commander, 63
      Incident Command Post, 64
      Incident Safety Officer, 63–64
      Liaison Officer, 64
      Public Information Officer, 64
   incident management, 68–77
      delegation of authority, 68–69
      incident action plan, 76–77
      incident communications, 74–76
      management by objectives, 71–72
      organizational flow chart, 69
      organizational principles, 69–70
      resource management, 72–74
   Logistics Section–Logistics Section Chief, 66
   Operations Section–Operations Section Chief, 64–65, 86
   organizational functions, 60–67, 69–70
   Planning Section–Planning Section Chief, 65–66
   Presidential Directive 5, 105
   purpose of, 59
   Unified Command, 60, 78
   Unified Command Structure, 68
National Interagency Fire Center (NIFC), 21
National Park Service, 165, 166
National Resource Typing, 106
National Wildfire Coordinating Group (NWCG)
   fire control strategies, 86
   fireline construction, 167
   Flagging Standards, 32
   flame length and fire intensity, 125
   Incident Response Pocket Guide (IRPG), 21
   PMS 506: *NWCG Interagency Single Engine Airtanker Operations Guide*, 178
   progressive hose lays, 117
   Red Card, 168
   S-219 *Firing Operations*, 181
   spotting zone, 137
   Staging Area, 119
   structure situations to avoid, 138
   tractor-plows, 168–169
   Wildland Fire Incident Management Field Guide, 21
Natural barriers
   control line, 115
   fire spread, 14
   as safety zone, 24, 33
NFPA® 291, *Recommended Practice for Fire Flow Testing and Marking of Hydrants*, 2012, 156
NFPA® 1001, firefighter certification, 168
NFPA® 1002, fire apparatus driver/operator certification, 168
NFPA® 1051, *Standard for Wildland Firefighting Professional Qualifications*, 40
NFPA® 1906, *Standard for Wildland Fire Apparatus*, 105
NFPA® 1971, structural gear standards, 35
NFPA® 1977, *Standard on Protective Clothing and Equipment for Wildland Fire Fighting* (2016), 36
NIFC (National Interagency Fire Center), 21
Nighttime (radiation) inversion, 10
NIMS-ICS. *See* National Incident Management System–Incident Command System (NIMS-ICS)
Nozzles
   ground sweep, 162
   remote control, 162
   uses for, 137
Nurse tender operations, 160
NWCG. *See* National Wildfire Coordinating Group (NWCG)

O

Objectives, incident (ICS 202 Form), 77
Offensive deployment of resources, 86
Off-road apparatus safety guidelines, 44-46
    capabilities and limitations, 45
    cautions during operations, 45-46
    hoseline safety guidelines, 45
    personnel transport, 46-47
Oil Spill Field Operations Guide, 71
One-lick method of line construction, 93-94
Operational leadership, 21-22
Operational period briefings, 76
Operational Planning "P," 71
Operations Section–Operations Section Chief, 64-65, 86
Organization
    Incident Action Plan (ICS 203 Form), 77
    levels of the Command organization, 62
    NIMS-ICS, 60-67, 69-70
Origin of a fire, 82, 98
Outside aid agreement, 73, 74

P

Parallel attack, 87, 98
Patrol, 108, 147, 150
Perimeter of a fire, 82, 98
Personal protective equipment (PPE)
    agency policy for maintenance and use of, 40
    boots, 36
    fire shelters, 37-40
    hard hats, 37
    leather gloves, 37
    self-contained breathing apparatus (SCBA), 33, 34
    structural gear, 35
    wildland gear, 36-40
        components, 36
        fire shelters, 37-40
        minimum requirements (NFPA® 1977), 36
        NFPA® 1971 standards, 35
Personnel. *See also* Crews; Teams
    command staff, 60, 62-64, 78
    NIMS-ICS organizational functions, 60-67
    transporting, 46-47
Pets, during ground cover/urban interface operations, 134
Phosphorous in fusees, 186
Pincer attack, 109, 150
PIO (Public Information Officer), 64, 148
Piped water systems, 155-157
Pits, hazards during fire operations, 30
Planning "P," 71
Planning Section–Planning Section Chief, 65-66
Plastic Sphere Dispenser Operator (PSDO), 188
Plastic Sphere Dispenser (PSD), 188
Pneumatic torch, 187
Ponds as water source, 157
Portable water tanks, 159
Power lines, 30, 53
PPE. *See* Personal protective equipment (PPE)
Precipitation
    defined, 9
    impact on fire behavior, 6, 9
Presidential Directive 5, National Incident Management System (NIMS), 105
Procurement Unit, 67
Progressive hose lays, 103-104, 117-118
Progressive line construction, 93-94
Propane torch, 187
PSD (Plastic Sphere Dispenser), 188
PSDO (Plastic Sphere Dispenser Operator), 188
Public Information Officer (PIO), 64, 148
Public relations
    dealing with the media, 148-149
    dealing with the public, 149
    operational program before the fire, 147, 148
    Public Information Officer, 148
Pulaski, 49, 171
Pump and roll attack, 105
Pumping capabilities of water tenders/ground tankers, 160-161

R

Radiation
    defined, 17
    fire behavior and, 13
    nighttime (radiation) inversion, 10
Radio
    communication during heavy equipment operations, 49-50
    lookout and crew leader communication, 32
Railroad bed shoulders, apparatus safety on, 46
Rain. *See* Precipitation
Rate of spread of fire, 3, 17, 125
Rear of a fire, 82
Recommended Practice for Fire Flow Testing and Marking of Hydrants, 2012 (NFPA® 291), 156
Red Flag Warnings, 23
Refusal of risk (turn down), 26-27, 55
Relative humidity
    defined, 8
    forms of, 8-9
    impact on fire behavior, 6
    temperature relationship to, 9
Resident shelter in place or evacuation, 121
Resources
    aid agreements, 73-74
    anticipating needs, 74
    deployment, 86
    ground cover/urban interface operations, 120
    impact on structure triage, 125-126
    National Resource Typing, 106
    national resource typing protocol, 74
    resource typing (NIMS-ICS), 105-108
    Resources Unit, 65
    single resources, 61, 73
    situational awareness concerning, 41
    structure and site preparation, 129-130
    support and operations, 165-178
        fire and aviation management, 174-178
        fire crews, 165-167
        hand tools, 170-174
        heavy equipment, 167-169
    terminology, 72-73
Respiratory protection, self-contained breathing apparatus (SCBA), 33, 34
Retardants, 144-145
Retreating, hit and run tactics, 142-143
Returning, hit and run tactics, 142-143
Ridges, 14
Ring firing, 145
Risk management
    defined, 22-23

Index   **205**

purpose of, 23
Standard Firefighting Orders, 23–26
turn down (refusal of risk), 26–27, 55
Road graders (maintainers), 48, 169
Rolling debris hazards during fire operations, 30
Roof
fighting roof fires, 138
impact on triage, 123
Rules of engagement, 40–43
lightning, avoiding, 42–43
look around, 42–43
look down, 42
look up, 41–42
situational awareness, 40–41
30/30 rule, 43
Rural water sources, 157

S
S-219 *Firing Operations*, 181, 184
Saddle, 12, 14
SAFECOM (aviation hazard) form, 27
SAFENET (ground hazard) form, 27
Safety
apparatus guidelines, 44–46
firefighters. *See* Firefighter safety and survival
hand tools, 49
heavy equipment operations, 48–51, 96–97
helicopter operations, 174–175
ICS 208 or 208H Form, 77
Incident Safety Officer (ISO), 63–64
mobile water operations, 158
Safety Officer, turn down (refusal of risk), 27
Safety zones
burned area as, 24
constructed, 33
defined, 32, 55
difficulty of escaping to, 30
engine cab as, 33–35
identification for safety, 24, 28
lookouts, communications, escape routes, and safety zones (LCES), 31–35
natural barriers, 24, 33
purpose of, 32–33
selection of, 33
structure as, 33
Sandvik, 171
Scalar structure of organization, 69
SCBA (self-contained breathing apparatus), 33, 34
Scene management, hazardous materials incidents, 52
Scraping tools, 171–172
SEAT (single-engine air tanker), 177–178
Section, NIMS-ICS organizational functions, 60
Section Chief, turn down (refusal of risk), 27
Section-level briefings, 76
Self-contained breathing apparatus (SCBA), 33, 34
Service Branch, 66
Shafts, hazards during fire operations, 30
Shelter in place, 121, 150
Siding, impact on triage, 123
Single Resource, NIMS-ICS organizational functions, 61
Single resources, 73
Single-engine air tanker (SEAT), 177–178
Site preparation, 128–130
Situation Unit, 65

Situational awareness, 40–41
Size of fuel, burning characteristics and, 5
Size-up
on arrival, 84
defined, 83, 98
before dispatch, 83
en route, 84
ground cover/urban interface operations, 119
incident priorities, 84
management by objectives, 71–72
watchout situations, 28
Sleeping near the fireline, 30
Slopes
fire behavior and, 13
impact on structure triage, 124
off-road apparatus operation safety, 45
slope winds, 7
Slopover, 83, 98
SMART acronym, 77, 85
Smokejumpers, 176–177
Snags, 54
Span of control
defined, 70, 78
Deputy Incident Commander, 63
Spot fires
defined, 98
firelines, 29
hazards, 82
Spotting, 14, 17
Spotting zone, 137
Sprinkler systems, 134
Stable atmosphere, 10
Staff. *See* Crews; Personnel; Teams
Staff-level briefings, 76
Staging Area
defined, 64–65, 78
ground cover/urban interface operations, 119
Staging Area Manager (STAM), 119
STAM (Staging Area Manager), 119
Standard Firefighting Orders, 23–26
Standard for Wildland Fire Apparatus (NFPA® 1906), 105
Standard for Wildland Firefighting Professional Qualifications (NFPA® 1051), 40
Standard on Protective Clothing and Equipment for Wildland Fire Fighting (2016) (NFPA® 1977), 36
Standpipes, 156
Strategic goals, 72
Strategic level of the Command organization, 62
Strategy. *See* Fire control strategies
Streams, apparatus safety near, 46
Strike Team
defined, 73
NIMS-ICS organizational functions, 61
Strike Team Leader at ground cover/urban interface operations, 120
Strip firing, 145
Structure
confronting the fire at the structure, 137–138
full containment, 138
no containment possible, 138
partial containment, 138
roof fires, 138
spotting zone, 137
exterior and interior preparations, 132–133

as fuel, 129
gas, turning off at the source, 133
hoselines, working, 135–136
porch light left on during operations, 133
pretreatment, 134, 141–142
protection of, 128–130
protection tactics, 134–135
as safety zone, 33
site preparation, 128–130
    clearance around structures, 130
    fuels, removing and trimming, 130
    on-site resources, 129–130
    structure considerations, 129
situations to avoid, 138
structure wrap, 134
triage. See Structure triage
Structure triage
    categories, 122
    defined, 122, 150
    factors affecting triage, 123–127
        fire behavior, 125
        firefighter safety, 126–127
        fuel, 124–125
        resources, 125–126
        structure, 123–124
    greatest potential threat, 122
    probable threat, 122–123
    structures that cannot be saved, 127
Subsidence inversion, 10–11
Subsurface fuels, 5
Supervision
    Division Supervisor, 120
    Group Supervisor, 120
    operational leadership, 22
Support Branch, 66
Support materials, 77
Surface fires, 4, 5
Surface fuels, 5
Survival of firefighters. See Firefighter safety and survival
Swamper, 36

T

Tactical level of the Command organization, 62
Tactical objectives, 72
Tactical worksheet, 76, 77
Tactics
    backfiring, 181–183
    burning out, 181, 183
    fire control. See Fire control tactics
    firing devices, 184–189
        drip torch, 184–185
        fusee, 185–186
        Helitorch, 188–189
        Plastic Sphere Dispenser (PSD), 188
        pneumatic torch, 187
        propane torch, 187
        Terra Torch, 187–188
        Very Pistol/FireQuick, 186–187
    hit and run, 142–143
    resource support and operations, 165–178
        fire and aviation management, 174–178
        fire crews, 165–167
        hand tools, 170–174
        heavy equipment, 167–169

    tender/tanker operations, 161–162
Tandem attack, 112–113
Task force
    defined, 73
    ground cover/urban interface operations, 120
    NIMS-ICS organizational functions, 61
    Task Force Leader, ground cover/urban interface operations, 120
Task level of the Command organization, 62
Teams. See also Crews
    strike team, 61, 73, 120
    Strike Team Leader, 120
Technical Specialist(s), 65
Temperature
    air temperature, 8
    causes of changes, 8
    defined, 8
    differences, impact on fire, 6
    fire behavior and, 6
    hot/cold dangers, 30
    relative humidity relationship to, 9
Terra Torch, 187–188
Terrain
    features impacting fire behavior, 12
    impact on fire behavior, 13–14
    off-road apparatus operation safety, 46
Thinking clearly during life-threatening situations, 25
30/30 rule, 43
Time Unit, 67
Tools
    brush hooks, 49, 171
    chain saws, 95–96
    combination, 171, 172
    cutting tools, 170–171
    fire-rake hoe, 171, 172
    firing devices, 184–189
        drip torch, 184–185
        fusee, 185–186
        Helitorch, 188–189
        Plastic Sphere Dispenser (PSD), 188
        pneumatic torch, 187
        propane torch, 187
        Terra Torch, 187–188
        Very Pistol/FireQuick, 186–187
    hand tools
        backpack pumps, 173–174
        carrying guidelines, 170
        cutting tools, 170–171
        for fighting ground cover fires, 170
        fire swatters (flails), 172–173
        safety, 49
        scraping tools, 171–172
        wire brooms, 173
    McLeod, 49, 171
    pulaski, 49, 171
    scraping tools, 171–172
    typical tool order, 94–95
Topography, 11–14
    aspect, 12–13
    defined, 11
    drainages, 12
    impact on structure triage, 125
    situational awareness concerning, 41
    slope, 13

steepness, 12
terrain, 13–14
Torching, 15, 17
Tractor-plows, 48, 168–169
Traffic control, 121–122
Training
    firefighter safety and survival, 165
    ICS-100, 60
    ICS-200, 60
    ICS-300, 60
    ICS-400, 60
Transporting personnel, 46–47
Tree hazards
    snags, 54
    unstable tree hazards, 30
Triage. *See* Structure triage
Turn down (refusing risk), 26–27, 55
Type
    protocol, 74
    resource typing, 105–108
Type 3 fire engine, 107
Type 4 fire engine, 107–108
Type 5 fire engine, 108
Type 6 fire engine, 108
Type 7 fire engine, 108
Typical tool order for fireline work, 94–95

U
Unified Command, 68, 78
Unified Command Structure, 60
Unit, NIMS-ICS organizational functions, 61
Unity of Command, 70
Unstable atmosphere, 11
Upslope winds, 7
Urban interface. *See* Ground cover/urban interface operations
U.S. Coast Guard, Oil Spill Field Operations Guide, 71
U.S. Forest Service (USFS)
    Common Denominators of Fire Behavior on Tragedy Fires, 43–44
    fire and aviation management, 174
    fire behavior characteristics resulting in fatalities or near misses, 42
    helicopter operations, 175
USDA Forest Service, 166–167
USFS. *See* U.S. Forest Service (USFS)
Utilities and heavy equipment operations, 50–51

V
Valley winds, 8
Vehicles. *See also* Fire apparatus
    mobile water operations, 158
    private vehicles for ground cover/urban interface operations, 133
    Vehicle/Machinery Operations (VMO), 158
Verbal Incident Action Plan, 77
Very Pistol, 186–187
VMO (Vehicle/Machinery Operations), 158
Volume of fuel, 6

W
Watchout situations, 27–30
Water
    application of, 139
    fire hydrants, 155–157
    mobile water operations, 158–162
        apparatus, 158
        ground sweep nozzles, 162
        nurse tender operations, 160
        portable water tanks, 159
        pumping and dumping, 160–161
        remote control nozzles, 162
        safe operation, driving, off-road use, 158
        tender/tanker operations, 161–162
        water shuttle operations, 159–160
    piped systems, 155–157
    to reduce heat buildup, 140
    resources for structure protection, 129
    rural water sources, 157
    supply at fire operations, 139
    water shuttle operations, 159–160
    water tender, 160–162
    wet lines, 144–145
    wetting down with, 139
Weather
    hot and dry fire hazards, 29
    impact on structure triage, 125
    precipitation
        defined, 9
        impact on fire behavior, 6, 9
    Red Flag Warnings, 23
    relative humidity
        defined, 8
        forms of, 8–9
        impact on fire behavior, 6
        temperature relationship to, 9
    relative humidity, impact on fire behavior, 6
    situational awareness concerning, 41
    spot weather request, 23
    Standard Firefighting Orders, 23–24
    temperature
        air temperature, 8
        causes of changes, 8
        defined, 8
        differences, impact on fire, 6
        fire behavior and, 6
        hot/cold dangers, 30
        relative humidity relationship to, 9
    watchout item, 28
    wind
        defined, 6
        direction, 7, 29
        downslope winds, 7
        general winds, 7
        impact on fire behavior, 6–8, 29
        local winds, 7–8
        Red Flag Warnings, 23
        slope winds, 7
        upslope winds, 7
        valley winds, 8
Websites
    FEMA training, 60
    National Interagency Fire Center (NIFC), 21
    NIMS Resource Management and Mutual Aid, 106
    resource typing (NIMS-ICS), 106
    S-219 *Firing Operations*, 181
Weeds, fires of, 3
Wet foam, 141
Wet lines, 144–145
Wide canyons, 14
Wildland fire. *See also* Ground cover/urban interface operations
    aerial photo, 4